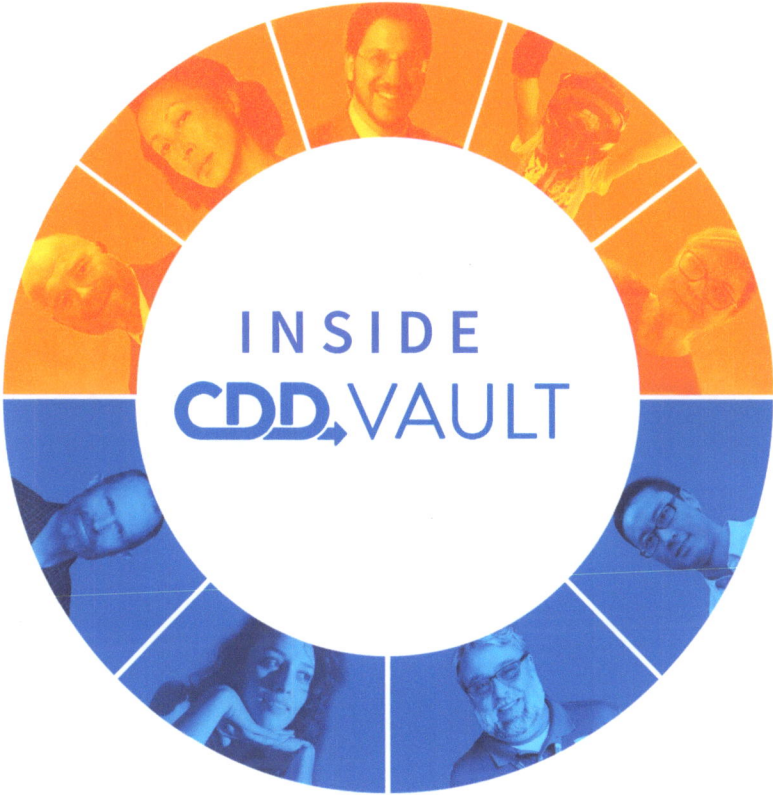

INSIDE
CDD.VAULT

INSIDE

CDD.VAULT

A Different Kind of Silicon Valley Success Story

BEHIND THE CODE:
The Human Side of Collaborative Drug Discovery

Published by CDD Vault

Collaborative Drug Discovery, Inc.
1633 Bayshore Hwy, Suite 342
Burlingame, CA 94010

ISBN (paperback): 979-8-9936443-0-1

ISBN (ebook): 979-8-9936443-1-8

Cover and interior design by Christian Storm | Stormhausen Design

First edition

Printed in the United States of America

10 9 8 7 6 5 4 3 2 1

This book is dedicated to
Dr. Alan and Ruth Bunin.

TESTIMONIALS

"RDKit and CDD have a symbiotic relationship with benefits across the scientific community. I really appreciate the pragmatic real-world expertise that the CDD team has brought to the work we've done together."

—**Greg Landrum**, PhD, "Benevolent Dictator for Life" RDKit

"Wow, I started reading and haven't been able to put it down. So much heart."

—**Ellen Berg**, PhD, Senior Scientific Leader in Human in Vitro Disease Models and CSO at Alto Predict

"At Medivation we used CDD Vault to collaborate 24/7 across three continents until it was sold to Pfizer in 2016."

—**Roopa Rai**, PhD, CEO Isostetix

"I remember when we founded Carmot Therapeutics and were trying to figure out how we would organize all the data we planned to generate. When I ran into Barry at a conference in 2010 and he told me about CDD I knew this would be a good solution. Thus began a productive partnership that led to several drugs and lasted until Carmot was acquired in 2024 by Roche."

—**Dan Erlanson**, PhD, Chief Innovation Officer Frontier Medicines, Author of popular Practical Fragments blog

"I remember brainstorming in the CDD offices. From my observation, the early start of CDD was very much a scientific and fun collaboration. The company name is totally appropriate."

—**Christopher Lipinski**, PhD of the ubiquitous drug discovery "Lipinski Rules of 5"

CONTENTS

INSIDE
CDD, VAULT

FOREWORD

Matthew Plunkett, PhD

Silicon Valley and the greater Bay Area are many things. It is ground zero for the global innovation economy: the birthplace of Google and Gilead; Netflix and NVIDIA; Oracle and OpenAI. A land of outsized one-syllable personalities: Jobs, Musk, Thiel, and Zuck. A complex mix of wealth and poverty, mushroom microdosing alongside fentanyl fatalities.

Silicon Valley began with Fairchild Semiconductor in 1957, where one of the founders coined the famous "Moore's Law" that accurately describes the relentless rate of progress producing continuously more affordable silicon chips, computers, the internet, and AI. The biotech industry began with Genentech in 1976. Its first product, recombinant insulin, was approved by the FDA in 1982; today half of all drugs are biologics. Both technologies revolutionized human health and economics.

CDD is right at the intersection of the two biggest Silicon Valley industries—software and biotechnology.

Many startups begin with lofty ambitions to change the world for the better. Google's "don't be evil" is perhaps the best known. Short-term horizons can erode an initially lofty vision. For many these "core values" posted in the entryway are derided as a joke by even their own employees. Yet for others, such words are words to live by. CDD is a rare example of a company that has walked the walk for two decades, through countless examples of challenge and success. CDD's Founder and CEO, Dr. Barry Bunin, continues to lead the company with its unique mission to enable impactful collaboration amongst scientists across the globe, within single companies and across different organizations.

CDD helps scientists around the world collaborate to discover new drugs for the rich and poor.

CDD upends the traditional hierarchy of corporate prioritization and values

CDD sees its customer needs as the north star, each of its employees as unique and valuable individuals, and business success as the natural outcome of getting the first two things right. The lessons here are relevant to any organization that is trying to make the world a better place. Individual stories of trauma, health challenges, abuse and death underscore CDD's mission to help scientists discover new medicines equally for all diseases—for maximum good for maximum people. It is easy to get numb to statistics, but not when it is personal, as described in the following pages.

Our backstory starts in India...

PART 1

AWARENESS

CHAPTER 1

Science Teacher—Barry Bunin, PhD

In the land of Krishna, on the road between Delhi and the Taj Mahal, the car swerved on the muddy street right into a tree. The driver had picked me up at 4 a.m. while it was still dark. I started the journey in the front seat but at some point, exhausted, I moved out of the passenger seat to nap horizontally across the backseats. I may have had a premonition. Suddenly, time seemed to be moving both faster and slower. My glasses flew off my face as my body twisted into the air, my ankle viciously hitting the front seats. I got my near-death experience as I awoke—all at once.

The spot where I'd been sitting was crumpled, front end kinked like a flimsy accordion.

The driver had been flung out of the car; he was moaning in pain, struggling to breathe. My best guess said broken ribs. A dump truck full of poverty-stricken people turned to stare at this roadside show. It wasn't long before two cabs saw an opportunity for a customer and stopped to help. We gently carried the driver into the back seat of one of the cabs. He told me to get his keys and his driver's license from the now mangled glove box. We moved my luggage from the trunk into the new car, and I hopped in the front, between the new driver and his friend or cousin; in India, with work so scarce, family members and friends were often brought along to help. There was a bobblehead of an Indian god on the dashboard, and they started to play some loud, traditional Indian music, "a yayayayaya," like you hear during the extravagant dance numbers in Bollywood movies. I said, "Hey! Can't you see he is hurt back there?" They said, it was for good luck.

No one died. You are in the land of Krishna.

True. We were okay. No one had died. And we were being protected on our journey. Krishna, responsible for both destruction and creation.

Visiting India, I saw poverty firsthand. The beggars with elephantiasis. The real problems people have. Compassion is the foundation, the raison d'etre for CDD, a more collaborative path to discover drugs simultaneously for diseases of the poorest poor and the richest rich. Nature doesn't care, so CDD supports both.

Our destiny was a small town in the middle of nowhere, a place too small for any map (I never did learn its name). Outside, on the hospital steps, sat beggars and lepers. The widespread poverty, the visually striking illness, was eye opening. I thought, here are people with real problems. We worry about our daily or weekly comparatively trivial soap opera events of normal life, and here are people fighting for their very survival. It has stayed with me to this day.

Everyone was calling me "Ambassador," as the only Caucasian in sight. We helped the injured driver into the makeshift hospital. They asked me to pay. I told them I was just helping the guy out and asked if I could use their phone to call another cab, since the current one was apparently just for short-distance trips. The hospital was too poor to have a phone, so they directed me to the one telegraph in town—a few kilometers in the direction they pointed. I limped off and found someone at the town's communications center, which wasn't much bigger than a phone booth for two. While waiting a few hours for the next cab, I found out the guy manning the town telegraph knew some chemistry, so we passed the time drawing some benzene molecules and telling the story of the serpent in Kekulé's dream.

In the famous dream of German chemist Friedrich August Kekulé, he saw a snake biting its own tail, forming a ring—an ancient symbol called the

Ouroborus. Basically, the snake represents the six electrons, in pairs, alternating bonds in a six-member hexagon benzene ring.

Which leads me to my even earlier origin story. At Mercer Island High School (Go Islanders!), Mr. Mills' Chemistry class had a student-teacher day—where the student got a chance to teach the class. My lesson was to teach the difference between a double bond (four electrons) and triple bond (six electrons). I found out the best way to learn is to teach. And there is great pleasure in seeing the light bulb go on in other people's heads when they get it. I liked the logic of chemistry. I sat in the back of the class with my Walkman on. Not exactly the class clown, but I liked to joke around. When it was time to teach, at first the other kids reacted by throwing crumpled up paper. After all, a fellow teenager wasn't exactly an authority figure. But as I began to explain how the extra electrons and their interactions with the nuclei of the carbons made the bonds shorter, and harder to break (homolytic cleavage) and it made sense in their brains, the classroom started to hush, to listen more closely ... and learn.

I was hooked. I wanted to be a teacher. A high school chemistry teacher.

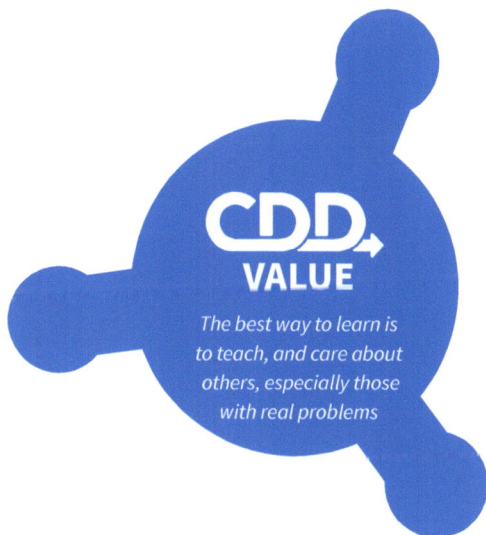

CDD.
VALUE

The best way to learn is to teach, and care about others, especially those with real problems

CHAPTER 2
Maximum Good for Maximum People

FOUNDER'S NOTE:

A for-profit company creates economic value. A non-profit organization creates social value. My father challenged me to create a for-profit that does social good in the world. And not just the type of good that every for-profit company does when it creates economic value, but an additional humanitarian value.

At Columbia University, I double majored in Education and Chemistry. I biked up to 135th Street in Harlem to student-teach at A. Philip Randolph High School. But then Columbia Professor Fine, who was responsible for educating the Chemistry majors, asked me if I wanted to do research over the summer on an NSF-REU (National Science Foundation Research Experiences for Undergraduates) scholarship. All I had to do was give him a reason. I replied, "because that's the only way I'll know if I want to do research."

I was placed in Professor Turro's Laboratory. I liked the culture, the logic, and being around super smart people all the time. My professor, Nick Turro, played handball with my college wrestling coach, Ron Russo, who taught me the mind gives up before the body, so strengthen your mind.

A postdoctoral fellow in the group, Dr. Robert (Bob) Rosenthal, took me under his wing. One thing Bob said to me and the other members of the research group foreshadowed the CDD ethos. Bob said, "If we had all the ideas in our heads in a single brain, then we would have the making of a Nobel Prize winning discovery."

The CDD Vault platform accelerates drug discovery for diseases of the poor and rich. The technical challenges are the same, so we simultaneously serve both markets via a superior web-based collaborative drug discovery platform. CDD board member Dr. Alpheus Bingham says, "making money isn't a mission statement, it's what society allows you to do when you create value." CDD's Mission is to provide an **unparalleled experience for humanitarian and commercial collaborative drug discovery.**

Thousands of years ago, the Buddha taught the ideal of maximum good for maximum people—a vow to help others. CDD Vault is a technical manifestation of this philosophical ideal.

"Collaborative" is the first word in the company name. From the beginning the vision was to allow any two or more scientists to collaborate. This allows for the economics of integrated specialization—chemists with biologists, predictions with experiments. Both types of expertise to iteratively discover safe, effective drugs—whether securely sharing results within a single company or between different organizations anywhere around the globe.

Catalyzing scientific collaboration for diseases of the rich or poor, has a multiplier effect. CDD provides value as a function of the number of scientists collaboratively exploring their important drug discovery data. The result is better medicines for patients discovered by leading biopharmaceutical researchers around the world.

Collaborative Drug Discovery balances the idealism of maximum scientific collaboration balanced with the realism that the majority of collaborative data must remain private. Therefore we named the platform "CDD Vault" to emphasize the security and privacy of a bank vault—even on a subconscious level. Customers are confident their private data remains private so that they can patent their discoveries.

The technical challenge is to help scientists arrange the atoms in the molecule of a drug organized just right to treat the disease, without harming the person. If your child is deadly sick, you don't care where the cure was discovered, you just want the scientists to work together as fast as possible on a cure before the disease takes your child.

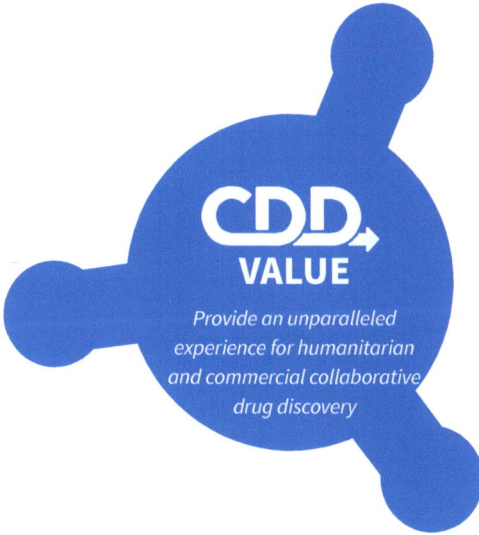

CDD. VALUE

Provide an unparalleled experience for humanitarian and commercial collaborative drug discovery

PART 2

TRAUMA

CHAPTER 3
Business is Personal—Kyle Riches

FOUNDER'S NOTE:

My high school wrestling coach's youngest son, Jay Jackson, wrestled at Stanford— and was a 2-time captain and 2-time NCAA Championship qualifier. After college, Jay was kidnapped by a deranged person with a gun yet still had the presence of mind and wrestling technique (while blindfolded, mind you) to capture his kidnapper, call 911, and get rescued by police. You can't control life, but you can control how you react. Wrestling and coaching legend Dan Gable was driven to greatness after his sister, Diane, was murdered. Dan moved into her bedroom to allow his family to move on after the tragedy, and used the deep pain to drive him to Olympic gold and to coach many other athletes to greatness. You can see the same thing with great artists like John Lennon, losing both his parents growing up, and channeling the pain into beautiful songs.

Kyle Riches has had to overcome a greater pain than most, losing his beloved 5-year-old son Bowie to meningitis.

At CDD, Kyle provides an authentic voice for our community, sets others up for success, is ultra considerate of other people, and hopefully the work helps him in his healing process.

One of CDD's mantras is "maximum good for maximum people"—for scientists in drug discovery to be of service for human health. Once upon a time, CDD had a program with Pew Trusts called SPARK (https://www.collaborativedrug. com/spark-data-downloads), where pharmaceutical giants like Novartis and Merck donated previous data generated in their search for new antibiotics. This data is hosted freely in CDD Public for the common good. When your own child is sick, you would do anything to help them.

When antibiotics work, they are usually only needed for a short time (no recurring revenues like statins which can be taken for life). And for antibiotic-resistant bacterial infections, they are prescribed even less often (less revenues), unless the antibiotic-resistent bacteria spreads. You personally hope to never need new antibiotics against resistant strains of bacteria, but you definitely want them available for the next Bowie, which could be your son.

◂

Here is the story taken directly from Kyle Riches LinkedIn post: "I rarely share anything personal on my LinkedIn or social media, but today is different.

Today, October 5th, 2023, is World Meningitis Day, and a few years ago, I was in the same boat as most of you are, likely not knowing there was a day dedicated to meningitis and its recognition.

But that has changed, as nearly 2 years ago, on January 20th, 2022, my wife and I lost our 5-year-old son to meningitis.

Our son, Bowie, was a gift to the world and everyone around him. He was kind, smart, loving, adventurous, and everything you could ever ask of a kid to be and then some. As we approach the upcoming holidays, I can't help but remember our last Thanksgiving. We had just moved to a new home and welcomed the entire family and extended family over to celebrate the holiday. Just before dinner, we all stood in a circle and went from person to person, sharing what we were most thankful for. After hearing a few family members share their thanks, Bowie caught on and asked if he could share. Bowie was jumping up and down with excitement to share and said, "I'm thankful for all of my people," and went on to name every single person in the room. This is a perfect representation of Bowie's priorities and what he valued in life.

Bowie made an impact on the lives of the people around him, whether you spent years with him or minutes. At Bowie's memorial, there were people in attendance that he had met while fishing with his grandpa, swimming at the pool, at community events, and at local restaurants, not to mention everyone who flew in from Washington, Oregon, California, North Carolina, and more. Bowie was a staple, so much so that there was an article written about him in the local paper about his presence in the community. People noticed the article, reached out to our family, and the memorial ended with standing room only. There were endless stories from nearly complete strangers who had met him along the way, which usually started off with Bowie asking them in his "adult voice," "So, how's your day going?"

The reason I share this today is twofold. Reason one is that meningitis is often overlooked, forgotten, or something that happens to "somebody else." It's something you heard of as a kid, or knew someone who knew someone who recovered. It's important that people know it's real and it's serious, and it's affected countless lives, ours included.

Number two is to continue sharing Bowie's heart with the world. Bowie's love towards others was unconditional and genuine; it didn't matter who you were, where you came from, or what you did. In honor and memory of my son, I now try my best to live the same way.

So, today I encourage everyone to be a bit more like Bowie. Give love unconditionally, be forgiving, approach the world and people with curiosity, and remember, above all else, the most important thing we have in this world is the people we get to do life with. Our people.

Here are some afterthoughts from Kyle Riches: CDD Vault has been instrumental in my healing. Part of that is because our work feels like it matters. Any one of the hundreds of teams we support might one day save another family from going through the kind of loss we did. And part of it is because I'm "thankful for all of my people" at CDD, who have been an ear, a shoulder, or just given me space when I needed it.

As you read this book, you'll see that everyone here is, above all else, human. A lot of companies talk about being "like a family," but to me, there's more value in simply being good humans. CDD is a company full of good humans who show up for the greater good, no matter what life has thrown at them, to help others move forward beyond illness and disease.

The CDD team has helped me keep moving forward. That personal support is, in many ways, a micro version of the macro impact of CDD: helping humanity heal, endure, and keep going.

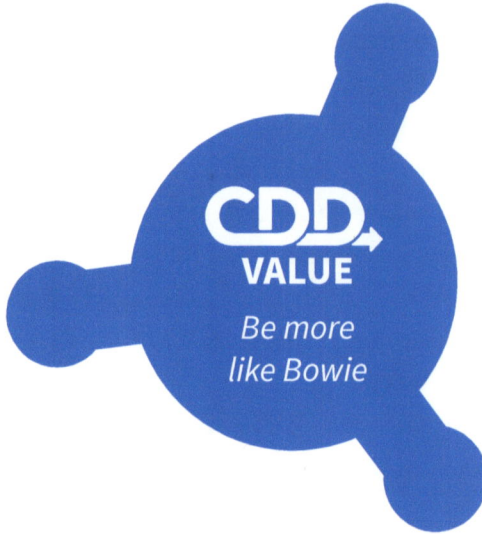

CDD.
VALUE

Be more
like Bowie

CHAPTER 4
Married with Children—Janice Darlington

FOUNDER'S NOTE:

Janice's story shows great strength and perseverance. She is now the CDD Voice... and married with children. The dream of CDD combines the idealism of every scientist collaborating together for better global health, balanced with the realism of knowing the world isn't fair, drug developers need to make profits, and pre-patent data must be private. Janice embodies that having overcome the deepest downs one learns to appreciate the biggest ups in life. I'm honored to work with someone so strong.

There is a powerful saying in Judaism from the Talmud, "If you save one life, it is as if you saved the whole world." Hopefully Janice bravely sharing her story conveys the hope that no matter how dismal a situation may be, it can (and sometimes does) turn for the better.

I was engaged twice before marrying my wonderful husband of nineteen years. Coincidentally, it was at the age of nineteen years that I met the first of three men who would propose to me. He was six years my senior and a graduate student in the lab where I learned to run my first PCR reaction.

I didn't dream of being in science. In fact, before my first science job, I ran away from the thought of having anything to do with biotech. My dad, posturing as a know-it-all, suggested I go into biotech. He said that's where the money was. I'm not sure he even knew what the term "biotech" meant or even its etymology. He was a self-proclaimed banker who emigrated to the United States from Singapore in 1972. He managed a couple fast food restaurants before turning to a life of driving taxicabs.

Growing up with parents who constantly fought and argued about money was immensely traumatizing and led to my altruistic view that money is the root of all evil. Even more traumatizing was the physical abuse I witnessed as a child when my dad lashed out at my mom and my sister.

The way I acquired my job in a marine biology lab at Scripps Institution of Oceanography during my sophomore year of college was fortuitous. I needed a second job as I alone was responsible for financing my college education. Through the combination of Pell Grants, Cal Grants, scholarships, and my part-time jobs, I was able to get by. Despite my dad's prompting to have college aspirations in light of an older sister who was a high school dropout, he had made no effort to pave the path for me. Never mind the countless $100 plate dinners my dad would bring me to show me off to his elite "club" friends. It didn't equate to money stashed away for me to go to college. In fact, on the morning of November 2, 1992, he dropped a huge bombshell on me as he drove me to school. Informing me that he had just voted in the general election, he said he wouldn't be able to pick me up from work or school from that day forward. He was leaving California and filing for bankruptcy. Destined for Nevada, he would continue taxicab driving. "Good luck with your mom," he said. "When she nags," he continued, "just stuff cotton balls in your ears." He left it at that, throwing me into a whirlwind of emotions, wondering what lay ahead. It was at that moment that I realized that the last four years had all been an illusion. It's something I should have sensed all along—that the life I was living was pretty messed up. It had been four years since my parents divorced following a restraining order filed against my dad after he relentlessly beat my mom senseless with a table tennis paddle. In the tiny quarters of our living room in our 850 sq. ft. condominium, he cornered her and beat her to a bloody pulp in front of my teenage cousin, my 5-year-old niece and me. I was blamed for the incident because she happened to be nagging me, when the violence started. My dad gave my mom no warning before escalating verbal sparring to physical sparring. But sparring would imply there was back and forth. My mom was all of 90 pounds when my dad attacked. While my mother was nagging me, my dad yelled "shut up or I'll beat you." She said, "go ahead," and there the relentless beating began. She found fault with me for not opening the door for her to escape. In the years that followed, my parents continued to live under the same roof until they didn't.

I was one of those kids who could have easily gotten lost in the school system. I was lucky enough to have a high school guidance counselor adamant about me going to college in the fall after I graduated and not taking that gap year I was contemplating. My friends and their parents, too, were my greatest cheerleaders in encouraging me to go. My entire education, in fact, had been a series of fortunate events directed by teachers, my friends, and my friends' parents, who were always looking out for me. They lived in the Palisades, Beverly Glen, Mar Vista, and Marina del Rey. I was the girl from "the wrong side of the tracks." To my mom, I was the reason for her misery. She was alone here in the United States after emigrating from Singapore. In Singapore, she was one of fourteen siblings. I was an accident. My mom would remind me time and time again that the day I was born, my dad was out picking up girls. My mom never wanted me to go to college because that would ensure that she had no one. With her not wanting me to go and the strained relationship we had, particularly during my teenage years, I felt going as far away as possible for college was what I wanted. I dreamed of snowy winters at Clark University or studying Communications at Boston College. In the end, my guilt led me to the decision to move far enough away, but not far enough that I wouldn't ever return, and so I followed my high school sweetheart to the University of California, San Diego (UCSD).

One thing I knew I loved was the ocean. The allure of working at Scripps Institution of Oceanography was strong. It sat steps away from the sandy beach where I could walk through the ocean waves. The lab on the top floor of Hubbs Hall had an unobstructed direct view of the Pacific. My resume was enticing to the hiring manager, a postdoc named David. He was a big fan of the TV Show "Married With Children," and of course, Christina Applegate, who played the role of Kelly Bundy, the promiscuous dim-witted daughter. If not for my high school internship at Mattel Toys and reporting to a real-life person named Kelly Bundy, things might have turned out differently.

The lab, for which I would get the opportunity to assist visiting postdocs from around the world, was one door down from the office of Dr. Andy Benson (famous for the Calvin-Benson Cycle associated with the 1961 Nobel Prize in Chemistry). Walking the campus of Scripps Institution of Oceanography on the way to the snack shack or to Friday TGs, I would cross paths with Dr. Kary Mullis, who was awarded the 1993 Nobel Prize in Chemistry for inventing the Polymerase Chain Reaction (PCR). And then there was the unique opportunity to travel to the Mediterranean Sea to be part of a research team working on the MEDEA program in addition to sitting in on graduate courses at SIO. All this felt exciting, but this also was probably on account of my intimate relationship with a graduate student.

It's not that I didn't love science or research prior to my job at SIO. I did, in fact, start to fall in love with the research process. My aversion to exploring the biotech route was only to spite my dad. I loved digging through facts and connecting thoughts. I became passionate about my research on malathion, controversial due to the Mediterranean fruit fly (Medfly) infestation in Los Angeles in 1989. It was my first endeavor into "research" for an 8th-grade research project. This first research project led me down the rabbit hole—from DDT (Dichlorodiphenyltrichloroethane) to the Vietnam War. It evolved to a growing fascination with history: the Vietnam War, John F. Kennedy, and Arthurian legend. Time travel would weave its way into my creative thoughts—the result of spending much of my impressionable teen years watching a TV show called "Quantum Leap," where the main character travels through his lifetime with a holographic companion, a former Vietnam POW, to "put right what once went wrong."

In my junior year of high school, the environment was very much on my mind, mostly because it was a global problem and by this time I was focused on International Relations. I wanted to change international policy, so when it came time for college, studying human civilization seemed like the right way to go. With that, I applied to Fifth College (later to be named Eleanor Roosevelt College). The college curriculum called for an intensive writing course. I apparently excelled at writing based on feedback from teaching assistants and my Making of the Modern World professors. I thought I was on course to become a writer. SIO altered that course.

My growing passion for marine microbial ecology was tied to my romantic relationship with this graduate student. Before I turned 21, we were living together. Before I knew it, I'd become the poster victim of domestic violence. Early on, R. (I've decided to refer to him as R.) showed signs of having a hot temper and being abusive. At the beginning of our relationship, he demanded that I completely cut ties with my high school sweetheart (the one I followed to college). When it was perceived that I hadn't, I found my head suddenly meeting the glass of his passenger side window. I immediately made my exit with no intention of turning back. It was always in my mind that I would never be like my mom and that I would never be the victim of a physically abusive relationship. And there I was, 4 years deep into one. I was exactly like my mom. That initial demonstration of physical force was followed by a week of profuse apologies and pleading for a second chance because, "of course," this was something he'd "never done before." Months went by and then it started again. By this time, I'd changed my stance: I wasn't going to be weak like my mom. I was smarter than my mom… "I can change him," I thought. We all know this story.

If you don't know how this story usually goes, I'll tell you. You can't change people, especially when they are unable to reflect on their faults. It was a flawed sense of hope, as my self-esteem and self-confidence quickly deteriorated. I'd become a prisoner in our home. The last of the people I severed ties with was my best friend, Lynn. I'd become a shell of a person, and she didn't know how to help. Everything came to a head in 1997 when R. spent months in Antarctica on a research project. This meant I was not under lock and key anymore. So when a note was passed along to me by my high school sweetheart's then current girlfriend to meet and talk, I took the risk to meet with him.

I can't recall why. Maybe it was because we never had closure. What I do know is that when all was said and done, R. would come to discover that I did meet with Jason while he was in Antarctica through a series of broken recovered messages—pieces of a puzzle he tried to piece back together when the hard drive of our computer crashed, and he worked to recover it. That's when I learned that deleted data doesn't necessarily mean it's deleted until it's written over. Of course, in putting the puzzle pieces together, R. made up the worst story possible. It all culminated with him not even being able to control his anger in light of whatever the criminal consequences were. Just before I made an attempt to take my own life, I tried to escape his uncontrollable fits of rage to control me. I tried to escape. I tried to run. He stopped me. He was bigger. He was stronger. He was out of control. He pushed me down and kicked me in the head. It was then that I decided that the only way out of the continued physical abuse was to take my own life. I locked myself in a bathroom and swallowed a bottle of his prescription pills.

Obviously, I'm here now. My suicide attempt wasn't successful. And I said goodbye to the idea of staying in marine microbiology. I was afraid of crossing paths with him again. Part of my success story lies in having acquired a paid internship at a small neuroscience company. There, I came across three headstrong women. Maybe I knew I was looking to escape, somehow I knew the only way out was to leave SIO. I call these women my three fairy-godmothers. If not for them, I wouldn't have had the courage to go to the police to file a report of the sustained abuse over the years. They took pictures of the "fingerprints" around my neck (the bruises from him wrapping his hands around my neck so tight), the bruises on my legs, the cut on my eye from when he knocked me down and kicked me in the face, and the bald spots on my head from where he would pull me by my hair.

Twenty-eight years later and I've been happily married for nineteen years. We have two daughters in their early teens with beautiful, kind souls. My high school sweetheart officiated our wedding ceremony with a chemistry book in hand. My husband's best man presently serves as the UCSD Department of Chemistry and Biochemistry Chair.

I recently heard Scott Galloway say on his podcast, "Love is giving witness to someone's life. To notice them and their lived experience." It's a thought that strongly resonates with me as I think about my kids, my friends, and the community that surrounds us. I've often thought about this with each person who has crossed my path. There's been so much shared life with my husband. We see one another, truly and honestly. I can't imagine anyone better to share this life with. As it turns out, I found him hidden in those memories that I sought so much to leave behind. . . he was the bearded woman in a coconut bra and grass skirt standing in front of the glass sliding doors at a Halloween party circa 1996. That's when I was still deep in the thralls of my disastrous relationship. That first Halloween party was hosted by a group of SIO grad students including Megan McArthur, now a retired NASA astronaut.

While we may have crossed paths in 1996, we didn't know the other existed until another Halloween, this time in 2002. This second Halloween party was hosted by my co-worker at Merck. Both my Merck co-worker and this mysterious Halloween guest who appeared every half-dozen years were both TSRI (The Scripps Research Institute) alumni. One thing led to another and as you may have guessed, the bearded woman in a coconut bra and grass skirt is now my husband, and a scientist, to boot.

And it should come as no surprise to find me here working for CDD Vault considering how intertwined my life is with science. I was a customer first and then a long-term member of the CDD team. But before joining CDD, I transitioned from basic to applied research, in drug discovery.

If I couldn't be a marine microbial ecologist, I figured I could still be a microbiologist, so I kept with that career path albeit for a minute. The biggest impact on my life that first job after college offered was exposure to the fresh air at Colorado's 14,000 ft. peaks. It's where I gained more clarity in my life and was inspired to prioritize my healing. I would eventually return to the place where my three fairy godmothers saved my life. By this time, the small neuroscience company in San Diego that I previously worked for had been acquired by Merck. I was enticed to take on a job programming the automation in the high throughput screening lab at what became Merck Research Laboratories.

This is where I fell in love with tech. This is where I fell in love with innovation, and this is where I fell in love with data, or perhaps it's more accurate to say, this is where I completely fell in love with the scientific method. In this capacity, I also repaired malfunctioning equipment and was challenged with solving a multitude of problems where I had to be creative and think outside the box. Recognition for my skills and efforts led to my recruitment to my next three companies, and it was the second of these three companies that introduced me to CDD Vault. I've now been employed with CDD Vault for almost 7 years. I recently transitioned to a role I would have never imagined filling in the days when I didn't have a voice—the Voice of CDD. Before I became the Voice, I had the honor of being a member of our technical support team for 6.5 years to support our users across the world under the guidance of Charlie Weatherall.

In my quest to make right what is wrong, and to just learn about people, to learn about life, and to find good people doing good things in the world, I stumbled upon something really special here at CDD. The rewarding interactions we have with our brilliant customers and the company culture I've experienced here offer daily reminders of the positive impact I might be having on the world. It certainly has made a positive impact on me.

Having been stifled, quieted, and isolated, "collaboration" proves to be a powerful theme in my world. When it comes to being married with children, collaboration is key to the success of how our family functions—the trust, the communication, the dedication, and the love. It's how our family thrives. I believe, too, that this is how CDD thrives.

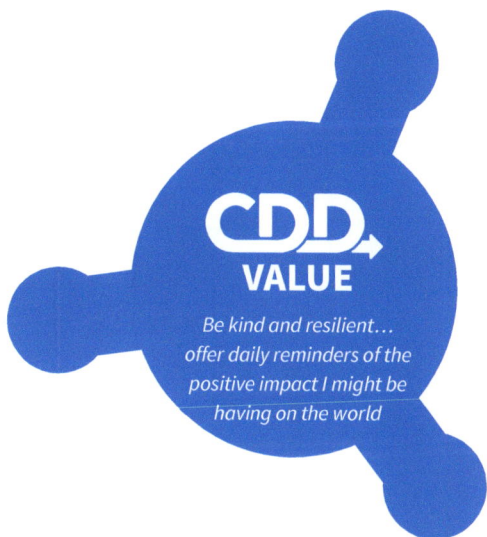

CDD
VALUE

*Be kind and resilient...
offer daily reminders of the
positive impact I might be
having on the world*

CHAPTER 5

Hearing the Kid Speak—Eric Putnam

FOUNDER'S NOTE:

Eric Putnam's personal story fits with the broader theme of overcoming adversity, of human health being a personal (individual) and global (universal) calling.

As with Janice Darlington, Eric worked in the laboratory and in technical support before joining CDD. At CDD he's part of the group we used to internally call "Team Awesome" and now call "The A-Team," harkening back to the popular 1980s TV series about outcast heroes helping ordinary folks.

You can see the empathy of our customer support team for our customers in Eric's story. Our customers on the Capterra software self-review website routinely provide 5 star rating, and one customer even gave our A-Team 13 out of 10 stars (on the 5 star rating). Here you can see why...

Growing up, I never wanted to be a teacher.

Which is funny when you consider that's *all* I've done for the last twenty years. While I didn't envision myself as a teacher, I *did* always know I wanted to be in science. I was thinking... marine biology. Instead, I stumbled into thirteen years of large-scale fermentation and drug manufacturing.

I worked on discovering ways to grow and propagate bacteria (and viruses) for new vaccines as part of a small Iowa company, which was later acquired by Pfizer. I became very familiar with the particulars of all aspects of drug manufacturing, including tearing down and rebuilding large fermenters, constructing growth media, and learning (sometimes by accident) how certain bacteria thrived.

I joined a project to oversee ways to grow an interesting little bug called Leishmania (the causative agent of Leishmaniasis). At the time, my supervisor's name was Lee, and we had resorted to calling this little bug "Lee's Mania" because nobody could figure out how to grow it in a large-scale setting. It was quite finicky and didn't grow aggressively or, well, at *all* in culture. I had thousands of hours of fermentation time under my belt and while, yes, it was tricky, it could be accomplished if one paid close attention to the growth curve and pH. While I didn't stay with this company long enough to see the drug through, it was nice to get something done that had yet to be accomplished by anyone else.

At this time in my life, I had a strong desire to be part of something bigger—to make more of an impact—I had been the company designated trainer for my department for some time and had worked with many people who had come and then gone on to other positions in other companies. One of those folks had recommended me for a teaching position with the US Army, and the timing couldn't have worked out better. My small (growing) family was looking to get out of Iowa and closer to relatives back west—we ended up moving to Utah for that job.

What could a fermentation scientist possibly offer in a teaching position for the US Army? An interesting question which resulted in an even more interesting job. I had stumbled into becoming part of the US WMD SME training team. (The Army is big on abbreviations.). I'll break that down for you: WMD is exactly what it sounds like—weapons of mass destruction. SME is what they call "subject matter experts." In short, I was to train various groups within the US military, as well as other allied forces, how dangerous bacteria could be turned into something to be feared on a national scale.

Using what I had learned during drug manufacturing, I was to train those groups how to identify signatures of large scale manufacturing, what was needed, and how to mitigate (defeat) those threats, including how to stay safe in the event of catastrophic release. This training was conducted in classrooms, laboratories, and even in real-world, full-scale mock ups of other nations' manufacturing facilities.

Imagine, if you will, a geeky scientist standing in the most remote parts of the desert, at midnight, waiting for a group of highly motivated individuals to arrive by helicopter, to practice "training" on such a facility. That was my life for nearly fifteen years. It was, in every sense of the word, amazing. It was a career filled with absolute bucket list items. C-10 gun runs? Been there. Suicide seat in a Blackhawk? Done it. Training the smartest and most elite groups the military has to offer? Check. Mourned the loss of members of those groups when they returned the next year down one soldier? Also (sigh) check. It was sobering, humbling. and my absolute honor to do that job for so long.

You can imagine my disappointment when I realized we could/should be doing so much more to prepare those teams for new and developing technologies. The gene editing technology of CRISPR was in full media explosion at this time, and we were prevented from teaching this technology despite student requests to do so. At one point, I concluded that I couldn't continue this career in good conscience knowing there was more I could be doing.

To be entirely fair, I had also missed so much of my children's growing up years, I felt I needed to be home more. Compounding that feeling was my beautiful third child. At three years of age she was diagnosed with NF1.

Neurofibromatosis type 1 is a genetic disease that causes tumors to grow along peripheral nerves in the body. Interestingly, NF1 affects every individual differently, some live their entire life never knowing they have the mutation.

My daughter was not so fortunate.

She is currently on her thirty-first MRI scan. We stopped counting the other X-rays and CTs long ago. She has a fantastic team of doctors who know her very well and monitor the tumors inside her body, the most concerning of those being the mass that is blocking her airway. When she was just four years old we were faced with the choice of whether to place a permanent breathing tube (tracheostomy), or to just. . . not. Which really didn't seem like a choice at the time, if I'm being honest. Some of you may know, but placing a trach comes with the additional side effect of removing a patient's voice. The doctors had told us before the surgery that she would likely not speak again without assistance.

One of my most traumatic memories is handing my four-year-old child over to the doctor who would place the tube that would allow her to breathe, and at the same time remove her voice. The idea of never being able to hear her say "Daddy!" again broke me in ways I can't express. And there I was, handing her over to the doctors, while she was terrified, screaming for her daddy.

In the days following the surgery we navigated new communication methods. My beautiful wife had the foresight to teach her some sign language ahead of the surgery, so we did have a limited ability to communicate some common phrases. I bought bells she could shake to get our attention when we weren't looking in her direction, but she didn't really like using those. So, true to her stubborn nature, she figured out how to talk again—day two post-surgery! I had come up to her room that evening after work and walked into her saying, "Hi, Dad." It was the most beautiful sound in the world. She made it a habit to scare her doctors when they would walk in her room and she'd do her best to greet them as they entered—none of them expected that!

As the years went by, I wanted to be more involved, however I could, in finding a treatment or cure for NF1. There I was, in the middle of the Utah desert, trying to figure out how to make that possible. The COVID pandemic was a true crystallization moment for me. It allowed me to focus on what I needed to do to be part of that. I felt so out of touch with real bench scientists by this point, like that part of my life had passed me by. How could a former fermentation scientist make any difference anymore? Bench life wasn't my game now; I was more used to teaching, and I enjoyed bringing that skill set with me.

How could I possibly combine the science and teaching fields into something that might make a difference? One day during the lockdowns it came to me. I knew a little about the field of scientific data management platforms, and I knew how disorganized a typical bench scientist's data could be. I knew what those hundred-page batch records looked like, and I felt there had to be a better way. I made the decision to go back to school and get my master's in technology management, with the goal of being part of making data collection easier somehow, to make better decisions and move science forward, if only a little.

I completely jumped out of my Army SME career and landed a role with a little company called Dotmatics. They were willing to take a risk on me despite my unfamiliarity with the data collection space. I learned a lot about what I liked in a data collection platform, what I didn't like—and more importantly: what worked for customers. My specialty was small biotechs; I could identify with their pains and offer a solution for those. As fate would have it, I joined Dotmatics during a period of upheaval, and my time there was short lived.

I still had a target in my mind but now felt like I was floundering a bit. Luckily for me, a friend I made during my time there saw that I was looking for a new place to land. That friend messaged me about an amazing company. I knew what I wanted, but I also knew what I didn't want. I needed to be able to move the needle to make a difference.

Enter CDD Vault and Charlie.

My first conversation with Charlie left me thinking that this man may have lost his marbles! There was no way a company could be spoken of this highly by someone who had been there so long (he had to know the dirt—where was the dirt?). All I was hearing was how wonderful it was. Something seemed off. And then I spoke to Janice, Salima, and finally Kellen. Independently, *all* said similar things. This can't be real, right?

Since that day, I have learned that the "act" is real. We meet twice a week with our product team to try to find new ways to make our scientists' lives easier. We have open lines to the CEO. Our developers *actually* care about how the product is being used, and not just about pushing out "features." I've never worked anywhere like this. When I have questions (and I've had many!) I can get the answers directly, and (often) from the person responsible for actually writing the code for that part of the platform. It truly is an amazing collaborative effort.

I've been lucky enough to be part of a support team, and I can't say enough about each of them. They all take a vested interest in the success of our scientist customers. They take meetings at odd hours, travel, and spend endless time preparing for meetings, just to push the envelope forward.

Even if my bench science days are long behind me, and perhaps I won't discover new ways of growing bacteria ever again, at CDD Vault I can help organize data collected by scientists to streamline decision-making and make that difference I crave. I can present them with solutions for all that disorganized data, to organize, share, and collaborate with others.

Oh, and the *real* kicker here? I'm currently working with a group of scientists to bring data into their CDD Vault, a group that is publishing data on NF1 drug therapies.

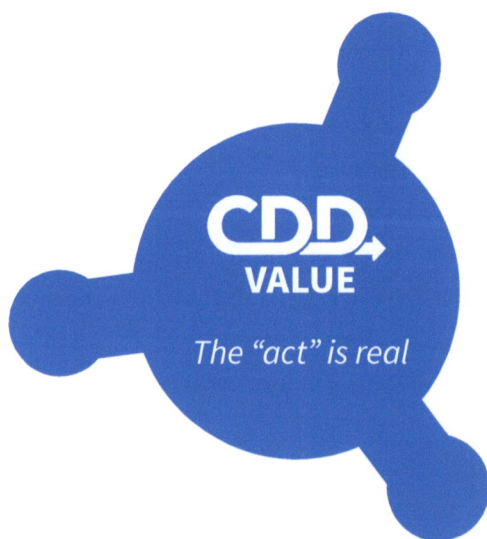

CDD.
VALUE

The "act" is real

PART 3

EMPATHY

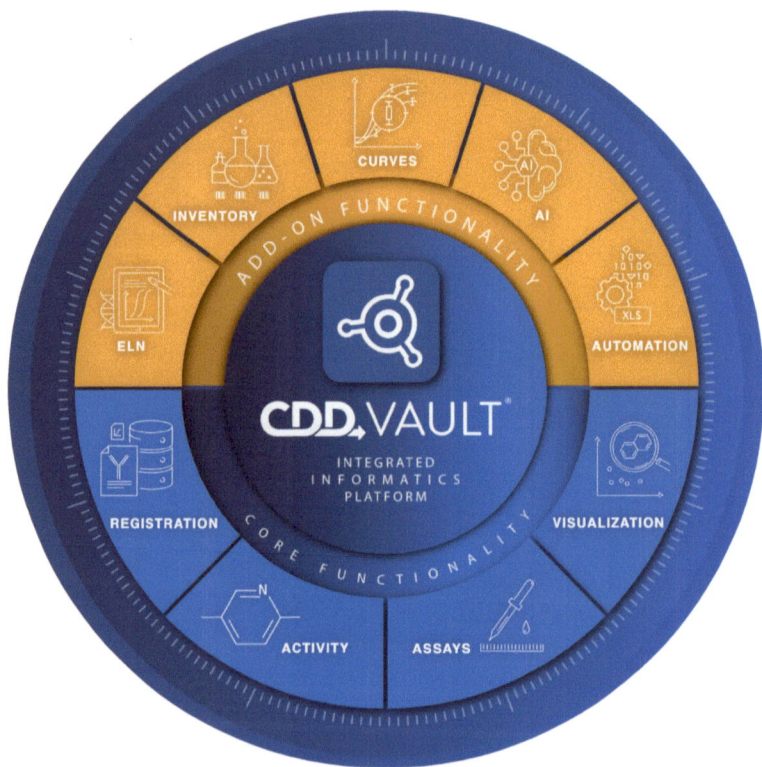

CDD hero image.

CHAPTER 6
Passion—Mariana Vaschetto, PhD

FOUNDER'S NOTE:

Dr. Mariana Vaschetto is responsible for optimizing our global operations and is a steward of our CDD culture.

Mariana conceived the CDD Vault graphic, also known as "hero image." An image so effective, clear, and powerful that at a glance one appreciates its meaning and value. Orange like a rising sun in an ocean of blue—a picture representing thousands of words, millions of lines of code, and billions of drug discovery data-points driving worldwide collaborative drug discovery.

At a company, you want to inspire people to work more, do more, to create value. Dr. Mariana Vaschetto is so motivated and self-driven that we often need to remind her to step back and take some personal time. It is a testament to her character.

Mariana helps us both feed ourselves and others. She leads our program to give back to any scientist working in Africa and simultaneously spearheads our international money-making side of the business— all with a great sense of humor and inspiration. Diseases like COVID can rapidly cross national borders as we saw during the pandemic. But ideas, data, and information via the web in CDD Vault can be securely shared faster. Nature can evolve fast, but humans have intelligence and the ability to aggregate intelligence. Mariana catalyzes global collaborations for human health.

There was this time in Korea when Kellan Gregory (CDD's Director of Product Excellence and Operations) and I were on our way to visit a customer, already running a bit late, of course. The building had these decorative fountains embedded in the floor, but you'd never know they were there unless you were really paying attention. . . which, clearly, I wasn't. One moment I was walking briskly, laptop and passport in hand, thinking through the upcoming discussion, and the next, I was in the fountain. Fully submerged. Laptop, passport, me, everything soaking wet. Kellan helped me out of the fountain. I couldn't stop laughing.

We didn't have time to go back to the hotel, so I did the customer visit anyway, dripping water and wrapped in the towels they kindly offered me. We still did the business meeting. The laptop, unfortunately, never worked again, but at least the customer got a good laugh, and they remembered us! Perhaps a video of me in the fountain is circulating Korea online or in some sort of TV bloopers program.

I grew up on a farm in Argentina, with an artist mother and a farmer father. My upbringing couldn't have been further from science; it was all about the land, creativity, and community. But in Argentina, diseases like Chagas, Dengue, Chikunguna, etc. are part of everyday life. Even though we weren't in an endemic area ourselves, I spent time with and loved a person who lived under the shadow of that *Damocles' sword that is Chagas*, a disease that quietly, yet profoundly, shapes your life. Eventually she died of Chagas. That early awareness stayed with me.

As I got older, my curiosity mixed, if I'm honest, with a little boredom. There wasn't much to do on a farm with a black-and-white TV and three channels! This boredom led me to discover science through some old books. When it came time to choose a path at university, I was torn between two passions: chemistry and social work. They seemed diametrically opposed, but I've always believed you can have more than one calling. I'd spent much of my teenage years volunteering and teaching children who needed extra support in underprivileged areas, so both fields felt meaningful.

I actually enrolled in both. After a year though, I realized the workload was too heavy to keep up, so I chose chemistry, partly because of my growing fascination with it, and partly because I had cousins my age studying pharmacy and biochemistry, which made the choice feel natural. There was something magical about chemistry, especially physical chemistry: the promise of endless learning, the freedom to work anywhere in the world, and the joy of understanding the invisible rules that govern our lives.

I was fortunate early on to meet wonderful mentors who opened doors for me and encouraged me to aim higher (though I also met some truly terrible ones, I catalogued them as my "lessons in resilience!"). During my PhD, I also met my husband, a brilliant chemist who has been my rock ever since. My journey eventually led me to quantum chemistry, where I stumbled into the fascinating world of designing new compounds, more math and computers, fewer smelly experiments. It was perfect for me. That intersection of technology and science completely captivated me.

After my second postdoctoral fellowship, I reached the familiar fork in the road that so many scientists face: stay in academia or move to industry. I was tired of the uncertainty and grind of the grant-work-grant cycle, so I made the leap to industry. I started in technical roles, which I loved, especially support (my first role) where you are always helping scientists to progress their quests. Over time, I transitioned into more marketing and business-oriented positions on an endless learning path.

But something was still missing: the human side, the caring side, the ability to give back and make a tangible difference, especially for communities in the Global South, where my heart has always been. There are so many gaps—in funding, in technology, in progress—that leave entire regions burdened by disease areas that are under-researched and under-funded simply because they are endemic south of the equator.

Then, in 2012, life threw me a devastating curveball: I was diagnosed with stage IIIb cancer (SCC), with metastasis in my lymph nodes. After a grueling treatment, I survived and came out with a post-treatment diagnosis of NED (no evidence of disease), and grateful to be alive. But it left me deeply aware that even in 2012, many cancer treatments were decades old (mine was good old cisplatin and radiotherapy). I thought: *We can do better. Technology should accelerate discoveries. It has to.*

A life-threatening event like that forces you to reflect on what really matters, what you want to leave behind. So, I left my job at the time. Through serendipity, and an old friend, CDD came into my life.

There were three reasons why joining CDD felt like coming home. First, the chance to work with an incredibly talented team. Second, the opportunity to assemble an international, diverse group of people as passionate as I am about science, technology, and humanity. And third, the ability to finally contribute to the field of neglected diseases, to bring all the threads of my journey together and put them to work for something that truly matters.

At CDD, I've found my place. It fulfills the deeper purpose I've carried with me all along: to help others through science.

With the support of Barry, Kellan, and many great colleagues, now real friends, I've had the privilege to assemble a brilliant, talented, caring, and collaborative team that helps CDD grow around the world, carrying the spirit of what makes CDD special. Everyone on the team brings their own perspective, shaped by their unique background and experiences, and that diversity is what fuels our creativity and growth. In our team, there are many different ways of seeing the world come together. Not everyone fits into a single mold, team members challenge assumptions knowing that they are heard and safe. That freedom, combined with a deep respect for our mission, allows people to bring their best ideas forward and feel a real sense of ownership over the impact they create.

The thread that connects all of us is a shared set of values: caring for our community, collaborating openly, and believing deeply that our work matters. When people feel safe to be themselves and know they are valued, that's when the most meaningful and innovative work happens.

I believe that I, and all of us at CDD, have a unique opportunity to make a difference beyond just serving our customers. I'm not here just to deliver

CDD Vault; I'm here to contribute to something bigger. And with that comes the obligation to act ethically, inclusively, and with care.

I take that responsibility seriously. For me, it means supporting research teams working on diseases that disproportionately affect the most vulnerable, often in underfunded and underserved regions. It means empowering scientists everywhere to collaborate and advance discovery. It also means fostering an internal culture that values people as much as results, making sure my own team feels supported and inspired to do their best work.

I'm proud to be part of an organization that is committed to closing gaps in access to technology, resources, and opportunity, while staying true to the values of caring and collaboration that define us. To me, that's what responsibility looks like: leaving the world a little better because I was here.

Over my years at CDD, I've been privileged to see that philosophy come to life through projects that embody the very best of what we stand for. Initiatives like the Gates Foundation TB Drug Accelerator, the COVID Moonshot, and the African CDD Vault program have shown me what's possible when talented, dedicated people come together with a shared purpose: to use science and technology to serve humanity. These have been some of the most fulfilling moments of my career, not just because of their scientific impact, but because they reflect the heart of why I do what I do. Helping to create, nurture, and deliver projects that have meaningful, lasting impact on people's lives is, to me, the most important way I can honor the responsibility I feel as a scientist, and a human being.

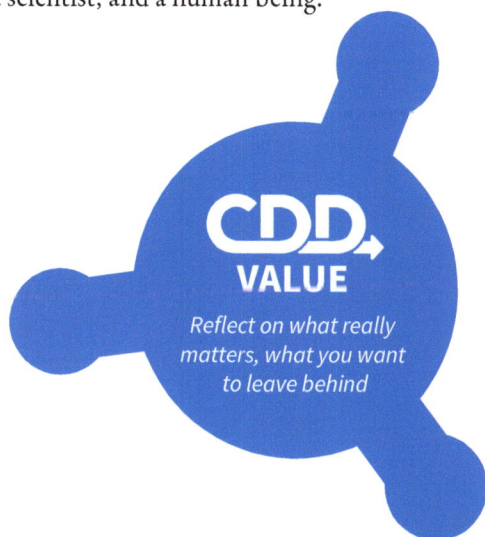

CDD
VALUE

Reflect on what really matters, what you want to leave behind

CHAPTER 7
Designer — John McCarthy

FOUNDER'S NOTE:

One of the ways a technology differentiates is via its brand. Great designers crystallize an effective message.

I'm grateful to John for our designs, including the rising sun hero image of the CDD Vault platform which he brought to life. Part of the collaboration concept includes the power of network effects, such as the extension of CDD's network from Mariana to John. You can see the importance of personal connection in John's story.

Over a decade ago, I met Dr. Mariana Vaschetto at a client meeting in Oxford. Even then, I knew I'd encountered someone remarkable

I'm not a scientist, an academic, or a researcher—I'm a graphic designer. So what am I doing contributing to a book like this?

That, in itself, says something important about CDD.

Abraham Wang (CDD's Marketing Manager) encouraged me to offer a perspective from the outside. As a long-time supplier, I have no scientific training, no PhD, not even a university degree. And yet, I've always felt welcome.

The fact that Abraham trusted a dyslexic graphic designer (who's tested his patience more than once!) speaks volumes about the culture at CDD. (In the early days of the relationship, I attached a 25-foot banner to our studio reading: "VISUALI**Z**ATION"—a cheeky reminder to the team to use American English and spare Abraham the grief.).

Ten years and two-hundred fifteen projects later—thank you, Mariana, and Abraham. I can honestly say CDD Vault is like no other client I've known. My role is small compared to their extraordinary impact, yet I've always felt seen and appreciated.

Of course, you might expect me to speak well of a company that helps pay my bills. But a few summers ago, during a very difficult time in my life, when I was caring for my aging mother, I wrote to Dr. Barry Bunin to explain that I feared I was becoming less reliable due to her endless appointments.

His reply, sent in confidence, changed the way I viewed my situation—and myself. It shifted how I care for my mum and how I show up as a son. More than anything, it gave me hope, perspective, and a sense that I wasn't alone.

I still revisit his words when I need to. They remind me why CDD is more than a client to me.

CDD Vault is a business, yes, but it's also a philosophy, a community, and for me, a cause that brings out the best in me.

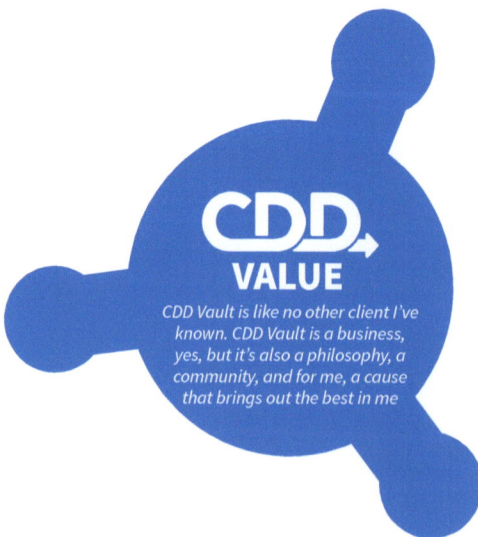

CDD.
VALUE

CDD Vault is like no other client I've known. CDD Vault is a business, yes, but it's also a philosophy, a community, and for me, a cause that brings out the best in me

CHAPTER 8

A-Team—Charlie Weatherall

FOUNDER'S NOTE:

Charlie Weatherall is the leader of our technical support "A-Team."

Our product team talks about product excellence, our dev team about dev excellence, and our support team ensures customer delight. They take it personally and Charlie most personally of all. Often in a support organization, members wait for someone else to pick up an email request from a customer (a ticket in industry parlance). At CDD the support team members rush to be the first to answer any question.

Our Zendesk ranking for responsiveness and customer delight is constantly in the 99th percentile (the top, not the bottom). In many companies support is often an afterthought (think being on hold at a call center). In contrast, at CDD we constantly put the A-Team on a pedestal and recognize "caring more" as a major differentiator. We also listen to every request from every customer in our weekly "market think" meeting—both to give immediate feedback and to prioritize the most important capabilities for the user community. After Dr. Peter Warner, our Gates Foundation Program Officer for our TB Drug Accelerator consortia, returned from a stint in Africa, we decided to implement a policy of providing CDD Vault software to scientists in Africa working on Neglected Tropical Diseases (NTD) free, the exact same as for our commercial clients.

Charlie Weatherall was so inspired that he wrote a guest blog from the heart announcing our policy:

CDD Announces Offer for Academic Organizations Throughout Africa to Use CDD Vault Without Charge

Collaboratively Managing, Analyzing and Sharing Data to Accelerate Neglected Disease Drug Discovery

Being an old timer in the life sciences/drug discovery software industry (I've officially passed the 30-year mark!), I am always surprised when something, well, surprises me. I've always known that Collaborative Drug Discovery (CDD) is a company with a heart. Our founder and CEO (Dr. Barry Bunin) always encourages everyone to explore ways to ensure that, as a corporate family, we make the world a better place. When I first joined CDD, this philosophy was even highlighted on the "corporate" slide-deck.

In the early years, a lot of this effort was based on helping researchers in the rare and neglected diseases space share their work. Personally, I've always taken great pride in CDD's involvement with the Bill & Melinda Gates Foundation's work with the Tuberculosis Drug Accelerator (TBDA). CDD

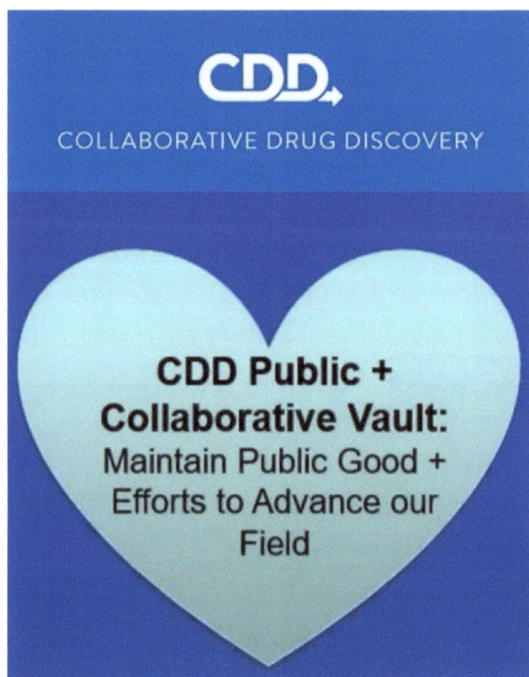

even initiated a program to provide donations to the Institute for Rare and Neglected Diseases. More recently, CDD quickly jumped in and supported efforts to find therapies for tackling the COVID pandemic. The overarching theme of these endeavors is the support of and encouragement to collaborate across the globe. Sharing is, in fact, caring ... and in my opinion, CDD cares more than most. In recognition of the important research and collaborations currently on-going in Africa, CDD continues the philosophy of giving back by announcing a policy for providing their flagship collaborative platform, CDD Vault, to scientists throughout the continent of Africa without charge.

ere is the rest of Charlie's personal backstory:
H
After obtaining my chemistry degree at a small liberal arts college in Mississippi, I worked in a research lab doing routine alcohol/drug testing for a state hospital as well as analysis, which was part of a March of Dimes research grant. It didn't take long for me to understand that the predictable, repetitious nature of lab work was not the best fit for me.

In 1987, I accepted a position as an information chemist at an international chemical manufacturing company based in Memphis. It was my job to manage the data generated by the chemists and biologists as well as query online scientific data repositories on behalf of the scientists. Note the date—in the late '80s, I was accessing mainframe computers via ethernet on amber monochrome monitors to perform text queries across online literature sources! Also, I find it interesting that Frank Brown is credited with introducing the term "cheminformatics" (or "chemoinformatics") in his 1998 publication while I was given that job title a decade earlier!

As an information chemist managing mountains of research data on a global level, I needed software. We chose to license the software from Molecular Design Limited (MDL), an outfit I consider to be the grandfather of all scientific data management platforms. I was mainly the project manager as we built a system for registering structures, storing chemical reactions, and reporting biological screening results ... although I did do a little bit of scripting along the way.

With this system deployed, the job (again) operated in a fairly perfunctory maintenance regimen. While attending a users' group meeting for MDL, I applied and interviewed for a technical support analyst position. I received a job offer and moved to California! This launched my career in the life sciences software vendor space, a career that's served me well for over two decades. My son was born during my tenure at MDL, and we felt it beneficial to move back to Memphis to be close to the grandparents–they make the most excellent babysitters! At this point, I tried to resign from MDL, but they asked if I could work remotely for six months as they looked for/hired/trained a replacement. I agreed, and they set up a ton of infrastructure to connect computers and route telephone calls from customers to my home in Memphis. I was, frankly, astonished when this six-month telecommuting trial turned into another six years!

After ten years helping scientists succeed in using the MDL suite of applications, I was contacted out of the blue by SciTegic, a start-up software vendor. They claimed a customer recommended me—I am forever grateful to that anonymous customer! I was intrigued by joining a small start-up and having a real chance at making a tangible difference for the company. I took the position as an application support consultant and began traveling the world to promote Pipeline Pilot. I knew I could do the end-user support part of the position but had never really worked with Sales in the pre-sales environment. This was my growth opportunity. A few years in, SciTegic merged with Accelrys (another large software vendor), and the small start-up culture, which I enjoyed, quickly evaporated.

A merger between MDL and Accelrys unexpectedly thrust my world into chaos. I felt that having twenty years of experience with both companies set me up for a great future in the new organization. However, the experience did not help as they sought to save dollars and down-size several senior folks, me included. I never would have thought that losing a J-O-B would affect me so deeply, but I was rocked. I knew I'd find another job, and I did. But the suddenness of an unexpected layoff was jolting. (To this day, well over a decade later, I have anxiety rooted in this experience.)

The bright side, of course, is landing at a company even better than the last and suddenly feeling like I got everything I ever wanted out of my career. CDD has truly done this for me. Towards the end of my tenure at Accelrys, I started receiving LinkedIn messages from Barry Bunin (CDD CEO). They weren't necessarily job offers or anything . . . just periodic "checking-in" messages. When he saw that I was in the job market, he asked why I hadn't reached out to him! Frankly, I was not sure he was even interested in hiring me, but I was honored to learn otherwise. Being so casual made it feel unique. I interviewed, and I'll never forget the words Kellan (my interviewer) said when he hired me, "You're our guy!"

Honestly, it felt more like I was being adopted than getting a job.

So here I am—thirteen years later leading a pre-/post-sales team of scientists from Hawaii to Finland! For decades, there's always been this "good-ole-boy" network of leaders setting up barriers. Barry and CDD are different. This is truly a collaborative company where everyone is valued, and the internal culture is given priority. There are no fake, artificial hierarchies to navigate; everyone in the company has direct access to anyone else. After I described what it's like at CDD to a candidate and what it's really like to work here, he said I made it sound like Willy Wonka and the Chocolate Factory. I like it! CDD is sweet but also a little quirky, unique. Shortly after hiring this candidate, I asked if I had misrepresented CDD in any way. He said no—he still feels like he's working at the chocolate factory!

The CDD team is encouraged to shine. The spotlight on the CDD culture helps ensure that everyone feels valued. The corporate mission statement dictates that employees are a priority. We collaborate internally just as hard as externally and this, I think, opens us up to creativity. There are no "my way or the highway" attitudes prevailing within CDD, and I often see peoples' minds start turning as they consider some new thought or idea that's been expressed.

I believe companies have a moral obligation to serve the greater community—and world—at large. I feel like a lot of companies don't do this. It can be a real competitive advantage to be known as the company who gives back, the company with a heart. But even that, highlighting this competitive advantage, almost negates the concept of greater good. Maybe the point is not to do good to gain some advantage/brownie points but instead to do good simply because it's the right thing to do.

Over the years, CDD has focused on neglected disease research, helping during times of crisis, and promoting research in neglected parts of the world. For example, as soon as COVID hit, CDD offered software and support to anyone doing collaborative COVID research for free. Several organizations took us up on the offer and collaborated within CDD Vault. One of my most meaningful memories was talking with a COVID researcher who had to run off to her next meeting, which was a weekly call with Dr. Fauci. (At the time, Dr Fauci was the director of the National Institute of Allergy and Infectious Diseases and the chief medical advisor to the President of the United States of America.) Little ole Mississippi me was actually helping a scientist who was also talking to a presidential scientific advisor on a weekly basis!

While I wear a lot of hats at CDD, the one I'd like to be remembered for is mentoring and looking after the A-Team. I do my best to cultivate an environment in which these diverse scientists can thrive and be happy. Nothing is as satisfying as receiving kudos from customers and colleagues. I am their biggest cheerleader, and that's the role I want to be remembered for.

It all boils down to relationships and respect. I never expect anyone to do something that I'm not willing to do. It's important to never forget that we are working with human beings who are all valuable in different ways. In fact, I often remind the team that it is OK to be human. Tell a joke, giggle with the customer, ask about their day, tell them about your kitty. I used to feel like a total cartoon character and wondered how on earth anyone took me seriously. I've learned, however, that my brand of silliness, coupled with a healthy dose of intelligence, gives me an edge, which makes me memorable and fosters an approachable personality. Remaining "professional" does not automatically mean you turn off your personality.

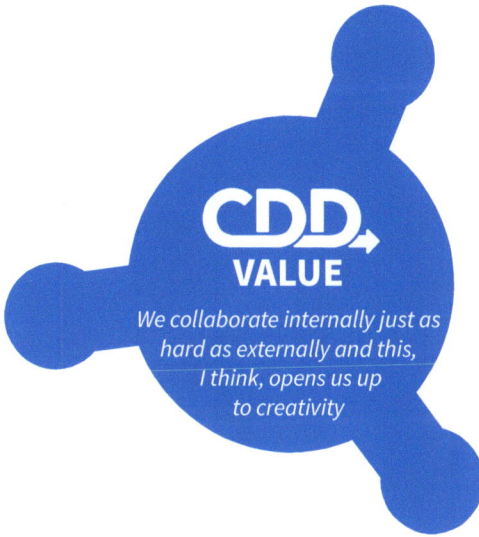

CDD.
VALUE

We collaborate internally just as hard as externally and this, I think, opens us up to creativity

PART 4

COLLABORATION

CHAPTER 9
Rubik's Philosophy

FOUNDER'S NOTE:

When all departments at CDD are perfectly aligned, complicated problems are solved like a Rubik's Cube. Hence, Rubik's is shorthand for our management team, where collaborative awareness manifests. At CDD, 'Rubik's Philosophy' means having great empathy for other departments (and people).

The famous Chicago comedy troupe The Second City teaches people to improvise, to riff off each other. "Yes, and" can take a good idea to great. The "Yes, and. . . " principle can be applied to business: Acceptance (the Yes part) + Contribution (the And part).

The founding CEO of Pixar, Ed Catmull, wrote a business book titled "Creativity Inc." where he refers to "plussing," taking a good idea to great. Challenging ideas, pressure testing storylines to reproducibly create top box office hits. One can create a culture for directed group serendipity—virtuous cycles instead of vicious cycles. Setting up the conditions for one happy accident to lead to a happier one. This is the collaborative ideal. A culture that can be engineered.

When I was at Berkeley in the Ellman Group, my labmate, Guangcheng (Kevin) Liu, gave an exhaustive talk on linkers for solid-phase synthesis. Listening to his presentation, I realized we had our hand on the pulse of the field with all the exploding publications. So while writing my PhD thesis, I also wrote a cookbook of sorts for chemists on library synthesis called "The Combinatorial Index." It is a useful book and became an academic bestseller amongst the small cohort of synthetic chemists looking to make more molecules faster and in parallel. And it got me thinking about all of the experiments done by all of the scientists in a specific field, rather than just the smaller numbers of experiments I could do myself. CDD has created a management structure behind a highly scalable software capable of capturing a wide range of experiments by diverse scientists, and to help them collaborate better together across disciplines.

⊰

The way individuals and departments choose to approach their work can have far-reaching consequences. If a person thinks only of themselves, pursuing their goals at the expense of others, they may still be able to accomplish what they want in the short term. One might optimize their processes in a way that looks efficient on paper but actually creates delays or bottlenecks for others. The short-sighted pursuit of metrics, isolated from the broader mission of the company, can end up making everyone else's work harder, more frustrating, and less effective. While such selfish strategies may produce local gains, they undermine the harmony of the organization as a whole.

A metaphor for this dynamic is the Rubik's Cube. At first glance, one might believe that solving just a single side of the cube is progress, and in isolation, it is. However, if the solver focuses exclusively on lining up just one colored side without regard for the others, the result can be disorder across the remaining sides. True success is achieved when every face of the cube is aligned, when the movements of one layer are coordinated with the needs of all the others. In the same way, a company operates at its best when every department is aligned, not when each department pursues its own perfection independently. Collaboration, awareness, and synchronization create a state where progress is collective and sustainable, not fragmented and temporary.

Thinking holistically within a company is so valuable. Entrepreneurs recognize that their success depends not just on their own effort, but on how well the entire ecosystem functions. When employees think like entrepreneurs, even if they do not technically own the business, they start caring deeply about group dynamics. They ask questions like: How does this decision affect the workflow of my colleagues? How does this process scale across teams? Will this create hidden costs for others later on? By shifting from an isolated perspective to a systemic one, individuals optimize for the collective dynamic. Instead of being a cog in the machine, everyone is an architect of an emergent, collaborative ecosystem.

At CDD the management team is called "Rubiks." Solving the Rubik's Cube means matching incentives, ensuring that everyone's success is tied to the success of the whole. At CDD, a unique profit and revenue-sharing program helps create this balance. Instead of employees feeling detached from the company's outcomes, they have real, tangible stakes in the results. The phrase "actions speak louder than words" becomes particularly relevant here. It is easy for leadership to preach values like teamwork, alignment, or customer-centricity. Rubik's provides a framework to walk the walk and hold each other accountable.

By giving everyone "skin in the game," CDD makes alignment more than just a moral or cultural expectation—it is also in everyone's self-interest. One of CDD's keys to scalability was creating this profit and revenue share plan. When revenue grows, everyone benefits. When costs are reduced through smarter collaboration, everyone feels the positive impact and shares in the profits. This structure reinforces the entrepreneurial mindset throughout the organization. Instead of department heads being motivated by narrow departmental goals and focused on politics, all employees are motivated by the shared outcome of the entire company.

Another value at CDD is the prioritization of stakeholders in an intentionally counterintuitive order: customers first, employees second, and the company itself third. At first, this may sound unusual—after all, many organizations talk about putting the company's interests first in order to ensure their survival. But the philosophy behind this hierarchy is both simple and powerful. By putting customers first, CDD ensures that the ultimate reason for the company's existence remains at the center of every decision. If customers are delighted, supported, and successful, they continue to do business with CDD, they advocate for the platform, and they help fuel growth.

Employees come second in this hierarchy, not because they are less important, but because their needs must be understood in the context of serving the customer. When employees know that their work has a clear and positive impact on the people they serve, their motivation deepens. They are not just working for a paycheck or to fulfill abstract corporate goals, they are working to create tangible value for others. And when employees are supported, respected, and rewarded for this work, they remain engaged and inspired.

Finally, the company itself, CDD as an entity, comes third. This may sound paradoxical, but in practice, it creates the healthiest long-term environment. When customers are thriving and employees are engaged, the company's success naturally follows. Profits, reputation, and sustainability are byproducts of prioritizing the right stakeholders in the right order. By inverting the traditional corporate hierarchy, CDD avoids the trap of focusing on short-term shareholder value at the expense of the very relationships that make the company viable long-term.

This approach also strengthens trust. Customers know that they are not simply being treated as sources of revenue, but as true partners. Employees know their wellbeing matters, and that leadership understands the importance of rewarding effort and initiative. And everyone knows that when the company succeeds, the benefits are widely shared rather than narrowly concentrated. This creates a culture where cooperation, rather than selfishness, is the dominant mode of behavior.

The Rubik's Cube metaphor illustrates the importance of coordination across departments, and CDD's profit-sharing program ensures that these ideals are reinforced with real incentives. The company solves the puzzle of organizational collaboration by aligning values, incentives, and outcomes for the benefit of everyone involved.

Entrepreneurship is a team sport. You can't just out-work everyone. You also need to out-execute the competition. That requires collaboration.

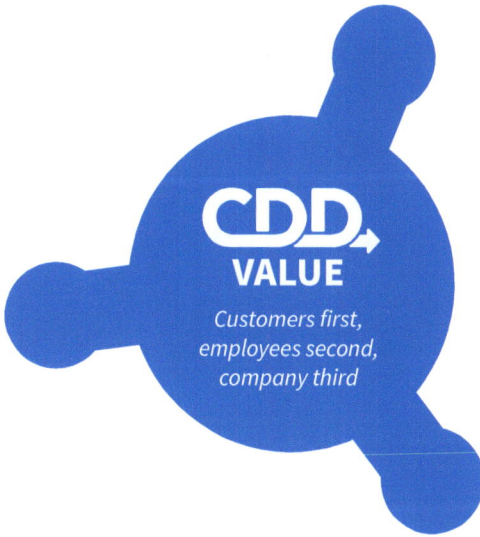

CHAPTER 10
The Sailor—Krishna Dole

FOUNDER'S NOTE:

If I get hit by a bus, Krishna's in charge of CDD. At least for a while.

As an engineer, Krishna uses minimal words. As a person, Krishna has rock solid values. Our original CDD Vault software architect said, "Krishna has an old soul." Typically engineers are thought of like Scotty in Star Trek "I can't change the laws of physics Captain!" and "I don't know if she can take any more, Captain!"

One of Krishna's many admirable characteristics is to see all the risks, yet remain calm in the eye of the storm. One of my favorite stories about Krishna is the time he was kayaking in the Pacific Ocean (one of his hobbies) and lo and behold, a shark swam right up to the kayak, yet he didn't flinch. Krishna is part of the dev foundation that helps CDD thrive through competitive shark-filled waters.

Although he loves kayaking, sailing, and fishing on the open ocean, we count our lucky blessings to have him right here on dry ground.

grew up on a meditation ashram near the town of Tomales, which is located in the ranch country surrounding Bodega Bay.

The ashram had a small community school, and I was fortunate to have some excellent teachers. Around my sophomore year they started to run out of curriculum and unofficially graduated us (even if I never received a high school diploma). At age sixteen, I started taking a full course load at Santa Rosa Junior College before transferring to UC Santa Cruz.

Growing up in the beautiful coastal prairie, we were raised with a lot of environmental awareness. This was also often a source of pain. As Aldo Leopold said, "One of the penalties of an ecological education is that one lives alone in a world of wounds." Since I enjoyed science, I picked environmental studies at Santa Cruz.

One of my professors strongly recommended a specialization to make up for the lack of depth of environmental studies, so I decided to double major in biology. Halfway through I discovered the Earth Sciences department. Of all the departments I'd encountered, they seemed to enjoy what they were doing the most, and since I was frustrated with the hand-wavey social science aspects of environmental studies, I switched to a combined earth science/ environmental studies major, keeping the biology major as well.

In college, the main role models you are exposed to are professors. I wasn't sure I liked what they had to endure. Talented graduate students had to be willing to move anywhere in the country for a job, regardless of where their spouse might end up. I loved science and published my undergraduate thesis in a peer-reviewed journal. But the thought of spending the rest of my life begging for grant money left me cold.

I noticed that many of the professors I worked with struggled with data management and analysis. I saw that the ability to write code could be a superpower, and started making slow, fumbling attempts to learn how to program. I can't take too much credit for the path my life has taken, but acting on this insight is what eventually led me to CDD.

I didn't take any computer science courses. I did, however, take a course in

Geographic Information Systems (computer-based mapping and spatial analysis), and ended up using those techniques for my undergraduate thesis.

The man who had been head of the GIS lab left to join a Silicon Valley startup called Metricom, which was trying to deliver high-speed wireless internet to all the metropolitan areas in the U.S. After I graduated, he hired me to help him automate spatial analysis and map generation for Metricom's wireless network. This was a thrilling time for me—getting exposed to RF engineering, GIS automation, the workings of a corporation, and software development. This was the year 2000, the peak of the dot-com heyday.

Unfortunately, Metricom did not have a compelling offering for the existing market at the time and burned through more than $600 million of investment before going bankrupt. I had remained through multiple rounds of layoffs and was there when the company was liquidated. Since I was living within my means, this was more of an exciting adventure than a calamity. I was offered another job but decided to take advantage of my independent and unencumbered situation to travel, visiting Thailand, Laos, Cambodia, and Nepal. Even with the international airfare, this proved less expensive than paying for rent and groceries domestically.

After these travels, I got a job as a post-graduate researcher at UC Davis doing programming for geomorphology simulations of the movement of river meanders. This is where I met my lovely wife, whom I followed while she pursued her PhD in entomology, first at Texas A&M, and then at Michigan State. Being far from Silicon Valley, these years were a challenge for my own career path. I seriously considered other ventures but ultimately had fun teaching myself to build software for academic labs. I was fortunate to work with Matt Yoder, who introduced me to Ruby on Rails, long before it reached version 1.0. We used Rails to build open source software for researchers doing phylogenetics and systematics. The software I worked on was used by multiple labs, including those studying bark beetles and parasitic wasps.

This is where I was when a recruiter contacted me about a role at CDD.

It was a great fit: a connection to biology and chemistry, a tech stack I was already using, and doing important work building tools similar to what I had been creating in an academic context. Except now this work would benefit from the resources and engineering discipline of the private sector.

CDD maintains a flat structure and minimal bureaucracy. Collaboration is a cornerstone of our culture: we solve problems together, learn from one another, share responsibility for outcomes, and reflect critically without assigning blame.

Like many, I want to work at a company whose core mission serves the greater good. CDD is focused on making the discovery of new medicines faster, easier, and less costly. We strive to align commercial success with scientific and humanitarian progress: by serving our customers in their research we benefit our employees, our shareholders, and patients everywhere, whether they suffer from common maladies or neglected tropical diseases.

I'd like to be remembered for two things: building something that made a difference and enjoying the journey with my teammates along the way. As a technical manager that means empathy and appreciation for the amazing people I work with, combined with passion for craft; I love contributing to software that does something useful.

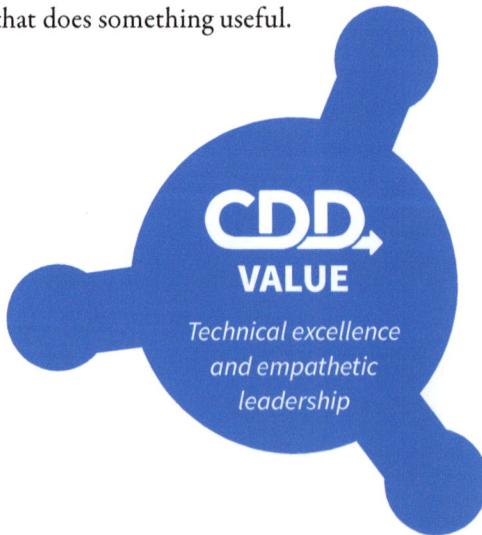

CDD.
VALUE

*Technical excellence
and empathetic
leadership*

CHAPTER 11
The Heartbeat—Kellan Gregory

FOUNDER'S NOTE:

Kellan is the heartbeat of CDD. Everyone knows he speaks the truth and genuinely wants to be good. He keeps a healthy skepticism and cares about every little detail. He appreciates teamwork over individual contribution.

CDD wouldn't be CDD without him. We take our work at CDD personally, for different reasons. Kellan represents the employee perspective, I represent the company perspective—and we both put the customer first, much like the Beatles did with their art. Whatever issues the Beatles had, whenever they pressed record in the studio, each member played each note and beat to optimally support what the song needed for the long-term legacy of the art. The same is true at CDD. When on the stage we are all about what's best for scientists around the world.

During my year off between Columbia and Berkeley, Virginia Cornish, a Chemistry undergraduate with me at Columbia who was now a Chemistry graduate student at Berkeley in Peter Schultz's group, kindly introduced me to Jonathan Ellman. All the other professors had fancy offices, yet Jonathan (still a postdoc in the Schultz group at this time) shared his ideas while doing High Performance Liquid Chromatography (HPLC) injections. Despite not having any office space, Jon's ideas were the most innovative (new and practical) and exactly what I was looking for. Instead of making molecules one at a time as had been done for ages, he wanted to make them in parallel. Libraries of hundreds or thousands of different molecules could potentially be produced all at once. It'd be like the Industrial Revolution finally hitting Chemistry a century later. I asked him one question, "Why hasn't anyone done this yet?"

"It's an idea whose time has come," he modestly replied.

While developing parallel synthesis in Jon Ellman's laboratory, I wrote a business plan. I was already getting the entrepreneurial bug but decided to stay in graduate school at Berkeley. Half a dozen other companies went public, surfing the combinatorial chemistry (combichem) synthesis wave until it became a common method for medicinal chemists. This entrepreneurial restlessness led to me later forming Libraria, which is where I met Kellan Gregory. I realized he was special, and reconnected when forming CDD to get the party started.

A s long as I can remember, I've always craved social interaction and felt a sense of ease when people and their sounds surrounded me. My dad would routinely leave music on (Frank Zappa, Violent Femmes, Pink Floyd, Talking Heads. . .) and when my mom asked, "Why?" he said it was so the rooms wouldn't get lonely. On a grade school trip to Joshua Tree, I tried clapping and noise-making during our hour-long solo expedition in hopes of getting a classmate to respond. Then, in sixth grade, my neighbor introduced me to the Friday Night Skate in San Francisco, a weekly procession of roller bladers traversing some of the city's richest and poorest neighborhoods for 12 miles, occupying the streets with as many as 1300 skaters. I was probably the youngest regular solo member of this loosely organized group on wheels, but informal mentors gradually emerged, including David Miles, the founder of the event. I jumped at the opportunities to make friends with all-walks-of-life people in their 20s, 30s, and 40s; some weeks, they invited me to a midnight meal at the Bagdad Café afterwards. I was a blader, and one close friend, Earl, a roller-skater who lived in a particularly tough stretch of the city, watched out for me on our weekly escapades. He was all smiles, quick to share life lessons and a contrarian perspective, which helped me develop a broader context for my private school education and social circles. Over the course of about three years, I missed only a handful of Friday night skates. Some of the societal values brought to life by this experience often came up again in the yearly SF Mime Troupe performances that my family never missed, perched on the hill in Dolores Park that overlooks the city skyline.

My parents prioritized education and world travel, often leveraging my dad's accrued miles from bi-weekly work trips for a three-week family international adventure. Destinations tended to be off the beaten path, highly researched, and rooted in culture or history. The days were filled with museums, architecture, top tourist attractions, wandering (getting lost!) and every interaction with local people we could wrangle. We were fortunate enough to be asked into private homes and move about with families or commuters on their buses and trains; in rural Thailand, we hitched rides on the backs of trucks for one spontaneous reason or another. Of course, we committed ourselves to the local cuisine. On some trips, we encountered people struggling through hunger, sickness, mental health crises, and extreme poverty (a town in Honduras comes to mind, where residents with next to nothing were routinely locked out of the consumer bank where we stopped to change money—utterly emblematic of sad, material social inequity). At the end of the day, my brother and I would join our parents to reflect and try to make sense of what we were seeing, inevitably imagining how different life would be if we lived here.

My parents supported my instinct to venture off on my own, and I made friends on these trips, usually kids around my age, locals and other travelers, at a pool, playing cards, filling a spot in pick-up volleyball games on foreign beaches. When languages didn't align, we relied on facial expressions and charades to communicate and craft an activity. On a flight home from Sharm El-Sheikh, Jordan and Israel, just before starting high school, I befriended a random 30-year-old man intrigued by the extent of my childhood itineraries; with a dozen hours in the air, we fell into a deep, accelerated dive of sharing key life stories that morphed into a telling moment of self-examination. I realized how much world travel and the flux of reflexive, momentary, memorable connections impacted my perspective, values and identity.

Customer engagement with CDD reminds me of connecting with people while traveling—appreciating new personalities, learning from a glimpse of their science, and forging a focused, purposeful relationship that can last. People who become our customers often care enough to carve out time to give feedback and share insights that help define the features we build. There can be chaotic moments, but with shared values and goals, there is so much value in teasing out the pain points with a scientist and then collaborating iteratively with CDD's acutely talented development team to implement more optimized solutions.

I mostly excelled at math and science. My mom, a clinical neuropsychologist with expertise in assessment (a data-intensive specialty) who took multidimensional calculus in college for fun, would lovingly feed me bonus math problems in middle school. The MCDS after school bus dropped my brother and me off a few blocks away from Tassajara bakery in Cole Valley; we'd grab a baguette and a table to do homework. An older man who often sat nearby noticed and he, too, began supplementing math problems with bonus breakdowns of some fundamental science concepts. But it was my inspiring 8th grade chemistry teacher who catalyzed my appreciation of science when principles first clicked, filling the subject with new meaning that followed me for a lifetime.

I was fortunate to experience two years of high school summer internships at SF General Hospital with world renown AIDS researcher Dr. Paul Volberding. I was tasked with recruiting subjects to a testosterone study (Anemia and HIV in the antiretroviral era: potential significance of testosterone). I was lingering in the phlebotomy lab on Ward 84, in the epicenter of HIV research, trying to convince patients to give a few extra vials of blood drawn in exchange for a McDonald's gift certificate. I had a short list of interview questions designed to prompt a detailed account of infection, treatment, and what it was like living with HIV/AIDS and related diseases. The next summer, I reformatted one of the key intake documents and had my first taste of data management and usability. Coincidentally, decades later, I met one of the Ward 84 nurses in a Vinyasa yoga class. We became friends, and over a few years of roughly 10 minute chats before each class, he pieced together a comprehensive backstory of the hospital and what it was like to live in the Castro district during the AIDS epidemic.

Informatics was mentioned on just one slide during my Tufts chemical engineering education. At the core of my biotech double major was a crash course in real-world processes taught by an adjunct professor from Wyeth who stressed the importance of data capture. That stuck, and even in the absence of details and methodology, in retrospect, it was enough to make an impression.

As my freshman year wound down, I was introduced to Barry's first company Libraria for a summer internship with Stephan Schurer. The long commute down to San Jose led me to negotiate four 10-hour days, the serendipitous reason I wound up interacting with Barry. It was a demanding project, I learned a lot about reactions and data curation, and most importantly, forged a strong relationship with Stephan. After graduating, I was ready to work, and it was an easy decision to return to SF given the biotech hotbeds and my ties to the city. I interviewed at a handful of startups for entry-level bench positions, never convinced that it was the right role for me. Everything was focused on cancer that year, such a meaningful domain, but without a PhD, it would be a long road and am I really a scientist? My mentor from Libraria Stephan Schurer suggested I reach out to Barry, who was just starting his second company. We met at a coffee shop in Noe Valley, and a few days later, I was a consultant at CDD.

My entire career has been with CDD, so naturally, the people I've encountered along the way have significantly influenced my personal, career, and domain growth in different ways. Elizabeth "Joey" Hansell was Jim McKerrow's lab manager and one of our very first customers. I would spend one or two days a week in the shiny new UCSF Mission Bay building, learning about their data (mostly trypanosomes), the drug discovery process, and the challenges of data management and supporting a team of scientists. Identifying "the Joey" person became a key success factor when onboarding new labs. Carolyn and Beth in Matt Bogyo's group at Stanford pushed our software to be better in those early days. Gathering feedback and meeting their data requirements using CDD Vault's early functionality was a fun problem-solving exercise. Our decision to focus on the long tail of drug discovery—academics in neglected disease research with a global outlook—made this whirlwind even more personally satisfying as a citizen of the world. Even beloved movies, like Medicine Man (seeing Phil Cruz's natural products at UC Santa Cruz) and Outbreak (visiting the BSL3 lab at NIH), were playing out in real life. I'm grateful for the experiences and opportunities to interact with thought leaders in our field who share their foundational understanding of drug discovery.

Internally, as our early prototype was morphing into a v.1 product, I was interacting routinely with our developers, starting to build the vocabulary to translate user requirements into specifications. One programmer, Mehran Bazargan, took the time to teach me the code structure and offer tips for better communication with his peers. We hired Moses Hohman to lead the development team; he nurtured the product and ultimately defined CDD's product principles while leading our transition to a contemporary development environment, all along pushing me to learn the latest UX principles. Sylvia Ernst instilled a business development foundation and shared insights into the history of informatics and our competitors. Maybe most importantly, she taught us what it's like to work at other informatics companies and how Barry is doing it differently.

Barry and I have a unique relationship. Sometimes brotherly, other times more distant, like preoccupied uncles, but always unfiltered. I did not know any better, so I never leaned into the usual boss-employee dynamic. Our compasses tend to point in the same direction, even if we'd draw different maps to some destinations. It was on a cross-country flight to DC that Barry and I got around to comparing life stories and discovering how much we have in common. We stayed with Barry's relatives, sharing a family meal and watching Jeopardy as we prepared for meetings at the NIH and DoD. I stayed with his parents in Seattle on a solo trip to visit customers of the Gates-sponsored TB drug accelerator program. We considered adopting some of Silicon Valley's unorthodox perks (a lavish office life) and outward-bound experiences (Burning Man) but firmly decided to stay true to who we are (spend within our means and volunteer over the holidays). We had some glorified startup moments, debating roadmap priorities at length over pizza and Mountain Dew in Barry's old apartment. Essentially, it is the trust we've built up over the decades that helps us operate and guide the organization as a living organism, awakening-increasing-containing-completing.

One of the most consistent interview questions I hear is, "Why have you stayed so long at CDD?" Our leadership staff have all been with the company now for at least 10 years. We all tend to insist on a work environment that sustains personal growth, opportunities, and respect. Barry encourages everyone to raise their hand and become involved. He can push people not only to accomplish more than they expect but to be skeptical about the so-called comfort zone. For me, this translates to years of challenging involvement in client support, sales, product development, marketing, accounting, and operations management with an undeniably exciting enterprise that explicitly strives to make a positive difference in the world.

I love the work itself—managing projects or teams, guiding interns or discussing problems, making presentations at industry events or in-house, addressing the myriad needs that arise within the organization, even debugging new releases. The CDD support staff is an inspiration—their persistence and dedication in finding solutions are amazing. Their mental agility is so impressive, switching between prospect calls, complicated training questions, and incoming support tickets, making every customer who calls on us feel important. I empathize with the challenges in each department. My baseline understanding of most facets of CDD and a desire to connect with others in a mutually satisfying way make me an easy ally. It's a vote of confidence when coworkers seek me out to vent frustration, and if I can't come up with an on-point perspective, at least they know I listen. I truly respect and admire these people. CDD is a smart, capable, collaborative team, and I genuinely look forward to working with them every day.

I channel these sentiments in a nuanced way when formally sharing our actual results versus quarterly goals. It's a difficult exercise that I embrace, taking the time to digest each department's progress and trying to summarize it in an impactful way. I've always been emotional at work, expressing my feelings and channeling what I'm hearing from others. In the early days, this was counterproductive, often leading to unnecessary arguments, mostly centered around ownership and control. My personal life, getting married and having two kids, certainly played a part in my evolution, but I learned enormously from the people at CDD. Particularly Krishna, who, true to his name, is so effective, exuding calm and thoughtfulness at every turn. He is a superb sounding board for all of us, with influence that ultimately makes us more productive. And I have learned to deploy my emotions for maximum lift internally and to generate excitement for our company's prospects and customers.

With refined processes for quickly assessing requests and mapping priorities (we call it Market Think), combined with Barry's talent for focusing predictively on the forefront, and our research informatics team's vision of the future, we share a high degree of confidence that we are focusing on what will matter long term. This is also reflected in our continued efforts to support neglected-disease researchers and empower consortia who are tackling the next big things. Certainly, there are elements outside of our control, but I'm convinced that CDD's explicit people-forward culture is a key ingredient in the company's secret sauce.

Legacy is such an overwhelming concept. I don't remotely aspire to have my name on a building or a park bench, and working at the top of a pyramid is more than I want (thankful for Barry). What I love is walking into a conference of industry veterans and being pegged, pleasantly enough, as young and inexperienced, palpably underestimated. Until I get to stand up and say I've been part of CDD for over 20 years (September 2005), before diving into the utility and benefits of the Vault, with references to well-known experts in our field whom I've been fortunate enough to meet over the years and citing a few enviable professional experiences. The impact of the emotional and intellectual shift in the room may be my legacy. But CDD Vault is becoming a household name, well-respected, and frequently known for the people supporting it. We are in the business of enabling and empowering scientists so they, in turn, can help as many people as possible. I am delighted and fulfilled to be part of that success, and a better person for it.

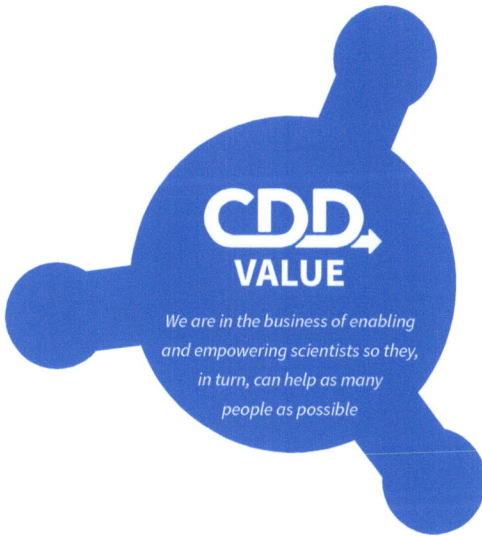

CDD.
VALUE

*We are in the business of enabling
and empowering scientists so they,
in turn, can help as many
people as possible*

CHAPTER 12
The Glue—Abe Wang

FOUNDER'S NOTE:

My wife was the first person who let me know Abe was smart and good. As usual, she was right.

People can be very emotional when they care, and selfish when focused on themselves. Abe, like Krishna, is a calming influence around driven people and agendas.

In many organizations there can be a rift between sales and marketing with each side blaming the other. Folks at sales like Abe in marketing. He has a biology background and enough experience and empathy to know what needs to be done, while being the glue between different people and views. Abe is a key piece of CDD's growth; CDD's growth coincided with Abe joining—whether correlation or causation, probably a bit of both, as is the case in any thoughtful collaboration.

I have always been envious of those who knew what they wanted to be when they grew up, because I was not one of them. When I graduated with a bachelor's degree in biology, I still had no idea what I wanted to do with my life. Given that I had some summer internship experience working in the lab, the path of least resistance was to find a job at the bench. I was fortunate to get hired as a research associate at SRI International, a nonprofit research institute with a long history of innovation. I started working in the molecular toxicology lab under Dr. Deborah Bunin. Debbie was a great first manager, who taught me not only good lab techniques but how to think critically. Under her mentorship I grew from a fresh graduate to a seasoned research associate who oversaw other junior staff members. Some of the interesting projects I participated in include looking for biomarkers that can predict drug toxicity, and the genetic effect of cosmic radiation on human cells. SRI used to host outside scientists to give seminars every Friday. On one particular day the guest was Dr. Barry Bunin, Debbie's husband. That was my first exposure to CDD. I was fascinated by what Barry described as a collaborative platform for drug discovery.

I remember walking up to Barry after the presentation, introducing myself, and asking him, "How do you market CDD Vault?"

He said that a large part of it was through "word of mouth," a testament to the fact that the Vault has enjoyed high customer satisfaction since its inception.

Eventually my wrist got tired of pipetting, and I figured I needed to do something different. After two years of part-time studying (while keeping my day job), I earned an MBA degree. Out of all the disciplines in business school, a career in product marketing seemed like the one to pursue. I saw marketing as a way of creating value—you can have the most amazing invention in the world but without marketing no one will know about it—which means no one will benefit from it. Changing careers is never easy, but when a small company gave me a chance, I managed to get a foot in the door.

Fast forward a little bit: I got a job as a marketing manager at Affymetrix, the company that pioneered microarray technology. There, I oversaw the gene expression microarrays, the exact product I'd used previously in the lab. So I went full circle from customer to vendor.

The day after I joined Affymetrix, the board of directors voted to sell the company to Thermo Fisher Scientific for $1.5 billion, so I immediately became a Thermo Fisher employee.

Thermo Fisher is a huge multinational corporation, with rigorous organizational structure and internal processes. I learned a lot of industry best practices during my tenure. I also got to meet and work with hundreds of talented people in the company (still a tiny fraction of the 70,000 headcounts they had at that time!).

One day, out of the blue, I got a message from Barry. This was years after our initial meeting. He said there was an opening in marketing at CDD. We started talking, and a few months later I found myself joining the CDD team.

Right away, I noticed the change in climate; CDD was the proverbial breath of fresh air compared to the corporate bureaucracy and red tape that I was used to. At CDD, I was able to get creative, I was able to make decisions quickly and execute on a strategy without jumping through hoops. People were incredibly friendly and collaborative. Of course nothing is perfect, but CDD really has a culture unlike any other organization I've been a part of; it allows employees to take on flexible roles, ones that maximize individual potential, while also helping foster contribution. Starting as a lone marketer, I am now part of a collective with others who share my passion and various expertise.

If I had to distill one lesson from this experience, it would be the importance of hiring the right people. Specifically, that cultural fit is more important than skillset.

Experience can be gained on the job, but a bad personality can be detrimental to the whole team's productivity and morale. We are blessed with great people at CDD, and I am thankful for each and every member who has contributed to our joint success.

It may be naïve to say this, but I believe marketers have a responsibility to uphold ethical standards. This is much easier to achieve when the brand is built on top of a solid product. Differentiation becomes natural if the product truly stands above competition. Storytelling feels authentic if the company genuinely stands behind its product and cares about its customers. And when customers become evangelists for your product, you have achieved the highest form of marketing: favorable, passionate word of mouth.

I am truly inspired by what the CDD team has achieved. Barry has found a way to preserve his original goal: to do good by humanity. Many founders set out to change the world only to get corrupted by the system. CDD has pioneered a business model that retains its independence and its culture, while being commercially sustainable for twenty-plus years.

We are lucky to be in drug discovery, one of the few businesses that truly has the potential to improve lives. Even among our peers, CDD stands out as a company with a heart.

At the end of my career, I will be proud of having contributed to a product that made the world a better place.

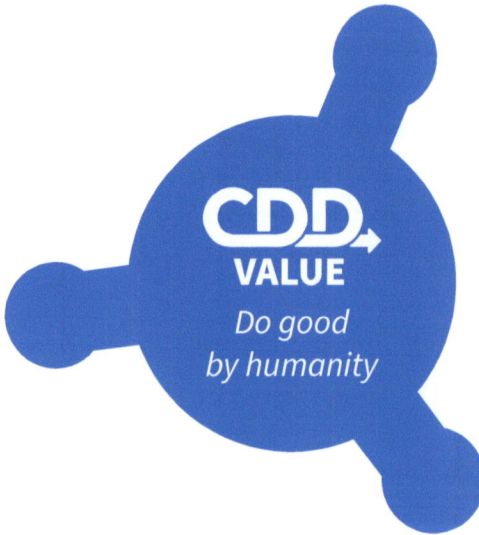

CDD
VALUE
Do good
by humanity

CHAPTER 13

The Oracle—Sylvia Ernst, PhD

FOUNDER'S NOTE:

As a founder, I like to control things. Control them to maximize the probability they turn out well.

In sales at CDD, we say the Account Owner orchestrates all aspects of the account—the plan, strategy, tactics, timing, communications, and when to bring in technical experts. Especially for selling, I like to control all aspects from a first good impression to the final close of the contract. So it was a big deal to recruit and trust Dr. Sylvia Ernst to set the foundation to scale CDD's customer base.

There are many things about Sylvia which have been crucial for CDD success, not just for her sales skills but for her disposition, her caring, her awareness of risk. Sylvia provides an early warning for anything we need to improve, before it gets bad.

Dr. Sylvia Ernst humbly says she's just a business executive, but she also won a prestigious award for the top PhD thesis using organometallic catalysts to split water molecules with sunlight to release energy (well before the modern environmental movement).

She was hired at Beilstein back in the days when reactions were on card catalogs and data was entered rigorously into the reaction database by hand. Somehow folks there found out she was charming (probably her genuine laughter), and they converted her to sales, where she became the company's top sales person. She's carried this through a long stint at MDL and a successful entrepreneurial experience at Scitegic (Pipeline Pilot), all before CDD.

It is hard to imagine someone who would complement the early team, while intuiting the future so accurately. That's why I call her "The Oracle" at CDD. Her second home is an organic orchard in Oregon, and when she's not tending to her fruit trees and vegetable plants, she keeps the CDD metaphorical garden healthy and thriving.

W hen I was eight or ten years old, my dad gave me an informatics tool kit, or "computer kit" as it was called back then, for my birthday (we're going deep in the last century for that one!). Computers did not really exist like today. There were not even pocket calculators yet. Phones had curled wires attached to them, and the TV was black and white with 3 channels. My geeky dad got me this kit from Phillips. I remember it was six or so blocks with little lights on them, a set of wires, and clips. A manual tutored the user by plugging in two wires, the second one being the Boolean operator "AND." Then three blocks for "OR," and so on. Half a year later, for Christmas, my grandma, probably nudged by my science loving mom, got me a "Cosmos chemistry kit," which started the chemistry experiments, which anecdotally led to thick white smoke billowing out from the basement. So early on chemistry and informatics were combined in that little girl's mind. Watching Star Trek, I dreamed, "This is how a computer should be! Not those little blocks and wires!"

If that was my start in the field, my first tangible experience was my first job. I was one of the dozens of writers for the *Beilstein Handbook of Organic Chemistry (Handbuch der Organischen Chemie)*. Still no computers. I received a wooden box, filled with paper slips and pencils. I then extracted data manually from literature onto more paper slips, before sorting them into wooden boxes. There was a dream. I so much wanted to be in the "computer department"! Finally, the chance came when they spun out the Beilstein Information Systems company, which was producing and selling an innovative product,

"Beilstein CrossFire" (now "Reaxys" by Elsevier), so I moved over there and into the freshly started sales department. Around 1994 for me, Chemistry, Informatics, and Sales all combined. A few smaller changes—like hopping continents and taking on new jobs—is what got me here today.

Collaborative Drug Discovery values culture and the individuals who make up the company. Barry Bunin, CDD CEO, has created a company that reflects his inner workings. Like Barry, we all want this to be about more than just doing business for money's sake. We aspire toward business with meaningful goals, and paramount in this is helping discover treatments to reduce suffering. CDD's environment nurtures collaboration over conflict. This in turn creates an atmosphere of trust where individuals can unfold and develop and unfold their talent. This is a unique ideology for a tech company. For me, coming from more conventional enterprises, it required time to adjust, but soon it felt like the most fun I ever had at a job. I hear that same sentiment from many others at CDD Vault.

We want a part in helping the world at large. One example is our continuous support of drug discovery projects for neglected disease research. For example: our current Africa initiative, where African academics get the full CDD Vault package for their labs for *free*. They receive the same quality, care, and support as our regular, paying customers, bells and whistles included. We are happy to provide our resources, a cost our company absorbs. Our sales team drives much of that. This is extremely rare in the world of sales, since a zero-dollar deal nets a zero-dollar commission. Their hearts are in it. Once you inform yourself about neglected diseases, the suffering they create, their incredible magnitude and threat to general world health, you have no choice but to feel good about this contribution.

The best I can wish for in my little life—not just my work life—is that what I give will have an ongoing, positive impact and weave into the constant streams of life. This is an underlying current to my existence: I want to be part of the good, enjoyable experiences, spiraling positive vibes, surfing the joyful side of life while inspiring others to ride this wave, too. Life seems very permanent; it is all pervading and a force which appears ongoing. It is just the coming and going of the matter it forms and which then disintegrates.

Even if memories are transitory, perhaps a few building blocks contributed will remain in the system, may it be in CDD Vault, or somewhere else (like when I help a trapped insect back to the outside of the house so it can go on with its life).

In the words of World War II hero Corrie Ten Boom, a Christian woman from the Netherlands who helped save nearly 1000 people from concentration camps: "Worrying does not empty tomorrow of its troubles . . . it empties today of its strength."

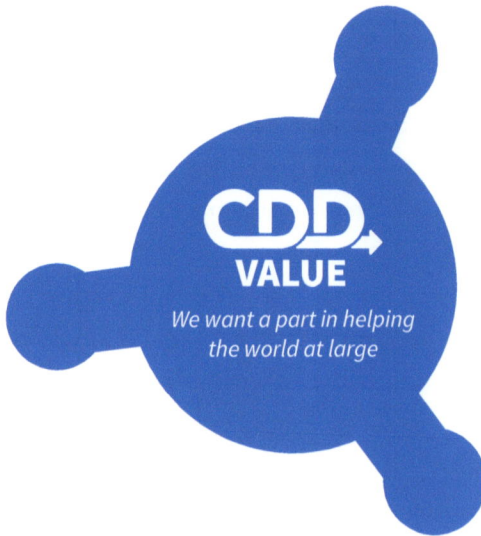

CDD.
VALUE

We want a part in helping the world at large

CHAPTER 14

The Tractor—Salima Ismayilova, PhD

FOUNDER'S NOTE:

I'm a bit in awe of Dr. Salima Ismayilova. Each new challenge we've given her, she has learned, mastered and become the teacher for the rest of us. Like Sylvia, Salima is a scientist and a farmer.

At first, she was nervous to present to others, but when our company's executive communications coach advised her not to think about herself but instead think about the audience, she was able to change her perspective 180 degrees.

It had been years since she gave a science talk, but we challenged her to give talks at the American Chemical Society. Hers became the CDD template for future talks. It isn't just her competence, but her rate of growth and fearlessness to take on new challenges. Every time she knocks a new challenge out of the park, we give her another one. Salima is now responsible for a larger group. I'm so glad she survived the black widow spider bite, and not just for CDD but for everyone who meets Salima.

was born by the Caspian Sea, Azerbaijan, as the Soviet empire was on the brink of a collapse. The years that followed were the most difficult, not just for my family, but for an entire nation learning to stand on its own. We had won our freedom, but we didn't yet know what to do with it, or how to shape a future that was finally ours to create.

Raising three girls in those chaotic years wasn't easy for my parents. My papa, a firm believer in education, always said, "My girls will stand firm on their feet; they will never lean on a man to walk through life." My mama would follow with, "It's easy to become a scholar, but even harder to be a decent soul." And I carried these platitudes as a pair of earrings my entire life.

After high school, I had no idea what I wanted to be, but I knew I needed an education. Some said doctor. Others said the teacher. But I chose agronomy, which was, admittedly, a puzzling pursuit in a nation that was so judgmental when it came to women and their careers. Perhaps it was because I've always loved food, not just the taste of it, but the journey it takes from field and farm to family. I wanted to be part of that magic.

I was one of only three women in my agronomy class. In an oil- and gas-rich country like Azerbaijan, agriculture was often overlooked. By my third year, the university shut down the entire department, and the path I had chosen came to an end. So I pivoted to chemistry and biology. While I was doing my PhD in Plant Biology, somewhere along the way, John Deere's local branch opened an unexpected door.

I defended my PhD thesis at twenty-six, the youngest PhD in the country, they said. After my defense, one of the professors suggested I slow down. He said, "Child, you're only twenty-six! What's the rush?" My response, "I promised my papa I'd beat him. He defended his thesis at twenty-seven." At the time, I was not yet grasping the joyful burden of being a scientist.

At John Deere, I was trained as a crop and soil scientist, before eventually moving into sales. I spent nine years living out of a suitcase, traveling the world, learning different farming techniques, being tolerant and accepting, and falling in love with soil again and again.

Those years were transformative. I witnessed the stark divide between abundance and scarcity, and I came to believe that if we could help shorten the distance between the two, even slightly, the world would be a better place for all. Traveling through low-income countries and witnessing the realities of food insecurity changed me. It made me cherish every breadcrumb on my plate. And somewhere in those moments of moving from one field to another, my first dream came into focus: I had to own a farm to grow crops.

Besides, how could I not fall in love with farming, when I was surrounded by green and yellow tractors everywhere I went? I was proud to work for John Deere, but slowly, working for JD lost its luster, and a sense of detachment became too strong to ignore. I'd become obsessed with numbers, and in the process lost touch with myself. It felt like my blood wasn't red anymore; it was green and yellow.

It was time to make changes. I left my family and well-paying job—I left behind everything familiar—and moved to the U.S. I knew I had to break free from the unspoken taboos and societal "norms" that were choking me. The move brought hope, but also many endless, lonely days. I was starting over in a new country with no family or friends at all.

The construction industry I landed in lacked empathy to begin with. There were very few women in construction back then, and the few who were there were not taken seriously, which added to the sense of isolation. Without mentors or examples to follow, it was harder to see a clear growth path to belonging. The job kept me materially afloat, but it pushed me away from my higher goal of contributing to human wellbeing. Ten years I lived without purpose, directionless, as if waiting for something or maybe someone to ground me.

In the summer of 2022, my nineteen-year-old niece Ayka passed away. I had known challenges and fear, but I'd never experienced grief like this. I didn't know how to cope with it. I did not know how to stop asking myself the same question over and over: How could I have prevented it? Four weeks later, CDD reached out with a sales opportunity. The timing was delicate. I hesitated. I was insecure about returning to science. Years had eroded the knowledge. The thought of starting something new while reconciling the loss of my young niece felt unbearable. How could I possibly survive the isolation of a home office when I could barely handle the loud voices in my head?

But my mother quoted someone: "Everything you want is on the other side of fear." So I accepted the challenge.

Since then, the journey has been fulfilling, full of growth, with hard lessons, wins, setbacks, healing, and now emerge with even deeper purpose and strength. Every day, I try to move closer to the four reasons I joined CDD:

1. Return to roots (science)
2. Surround myself with quality people (intellectuals)
3. Play big (serve humanity, however direct or indirectly)
4. And most of all, participate in delivering a cure (for the memory of Ayka)

CDD is a unique company. The atmosphere is full of warmth and authenticity, which is rare in the corporate world. At CDD, customers and employees are more than numbers; there's room for humanity. Barry leads with a blend of transparency, respect and fairness that sets the tone for us all. He challenges us, sharpens our thinking, and in doing so fosters real growth, both as individuals and as a team.

In this supportive environment, something began to shift in me. Faith and encouragement I kept receiving in the last three years helped me piece back together a confidence I had lost during the chapter of my life as a new immigrant with an "uncommon accent and imperfect English." As I dove deeper into this industry, learning about the immense suffering caused by rare and common diseases, something unexpected has happened: the sharp pain of losing Ayka is softening; grief is transforming into something I can carry with grace for the rest of my life.

At CDD, we may not cure diseases directly, but we contribute to the end goal. We are a bridge between science and patients. Because science isn't just technical or clinical. It's a philosophical belief that all life is worth fighting for. I'm glad to be part of that legacy.

CDD is home. It's where I belong now. Each interaction with colleagues and customers makes me want to be a better person. Although I am not an active scientist anymore, I help to enable brilliance in others (even if my contribution is just a drop in the oceanic petri dish). When I look back, I see clearly that I joined CDD for Ayka—she could not be saved—but through CDD we help scientists to save someone else's Ayka.

While I find joy in my work at CDD, I haven't let go of my farming dream. Today healthy cows graze on ten acres of my land, feeding several families (and yes, I have a brand new 45hp JD tractor). Life taught me that reaching your dreams requires hard work, resilience, dignity and perhaps most of all faith in the power of self.

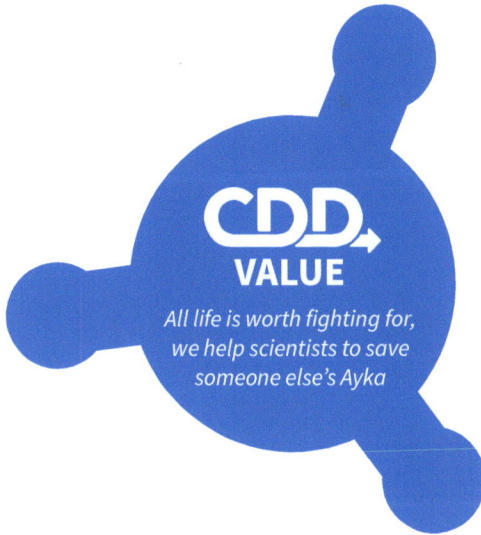

CDD.
VALUE

All life is worth fighting for, we help scientists to save someone else's Ayka

PART 5

INNOVATION

CHAPTER 15

Star Trek—Jacob Bloom, PhD

FOUNDER'S NOTE:

Jacob is hilarious. And not when he is trying to be. It is the rest of the time. But he's not just someone who lightens the mood with his laughter; he's also a foundational part of CDD's success.

Krishna trusts Jacob with the most important, complicated, and critical parts of our infrastructure. . . which hopefully means Krishna sleeps better at night.

We started CDD before AWS existed. We pioneered the hosted drug discovery concept with the largest server hardware metal that money could buy. We are now on AWS, but back when we ran our own servers, Jacob was the one to go into the facility in the middle of the night (or through rush hour traffic).

Many of our customers might not know what happens behind the scenes, but CDD's software development and product teams are very much aware that the "taken for granted reliability" doesn't happen by itself. It happens because of Jacob and our other super reliable technologists.

Nirvana is when everything Jacob Bloom and the development team does isn't even noticed, customers enjoy >99.9% uptime of the software application, without even thinking about it twice.

I have always been an inquisitive person. From a young age, I explored the world around me, never satisfied with simply asking what, always looking for the why and how. At seven years old I asked my mother, "What's the meaning of life?" Much to her credit, instead of shrugging it off, she wisely replied, "Be happy and make the world a better place." Since then, I have always put those goals at the forefront of my life. If things didn't make me happy, I would find something that did. If I wasn't helping those around me, I would look for ways to do so.

HELPING OTHERS BECAME MY PASSION.

I developed a yearning for math problems. The logic and rules made it easy for me to understand the why and how. It felt good to be able to solve these puzzles in an orderly environment. My father, an avid software and hardware engineer, guided my pursuits in a similar fashion. Rightfully recognizing the overlap between solving math problems and programming, he gave me a copy of Visual Studio. I realized quickly that I could spend hours in front of a computer screen this way.

PROGRAMMING BECAME MY PASSION.

Then came high school chemistry. I really liked chemistry. The rules and balancing of reactions made it feel like math. But there was something that irked me greatly; high school chemistry mostly taught the what, but not the how and why. Despite these feelings of incomplete knowledge, I had found the thing that would let me "Be happy and make the world a better place." I conveyed this quite inelegantly once, when asked what I wanted to do for a living, "Drugs. Make drugs. Uh, the legal kind." Not my best prose, but it's great for a laugh.

CHEMISTRY BECAME MY PASSION.

College is where I finally completed my knowledge of the why and how of chemistry, focusing my studies on quantum mechanics. My very first thought when I saw the spherical harmonic solution to the hydrogen atom was, "Wait, chemistry is all just math?" I fell in love with quantum mechanics and its ability to explain the chemical world through math. This exploration soon turned towards computational chemistry during a summer research opportunity. I soon found myself weaving programming into my favorite science.

COMPUTATIONAL CHEMISTRY BECAME MY PASSION.

Throughout college and graduate school, I thoroughly enjoyed being a teaching assistant. Helping others understand the concepts that I found so fascinating was truly fulfilling. As one of the longest-tenured developers at CDD, I try to mentor as much as time permits, passing on the knowledge I have gained. At CDD, we strive to put people first, not just in helping scientists develop treatments and cures, but in how we treat our customers and employees. I am encouraged to take time from my busy day to help my colleagues puzzle through their work. I even get to interact with customers so I can learn how they use our software, learn of possible features to add, and help them interface with our API more effectively.

HELPING OTHERS IS STILL MY TRUE PASSION.

Something I always tell prospective developers: the Collaborative in Collaborative Drug Discovery isn't simply a part of our company name, but our main value. For collaborations to work, everyone involved must feel valued, respected, and informed. This has proven vital for the development team throughout our expected and unexpected challenges. When COVID surprised the world, the team switched to working remotely without any loss in this collaborative spirit. Pair programming, mentoring, and group analysis have continued despite the now-normalized virtual nature of our communication. When we needed to swap out the chemistry engine that Vault relies on, we were given a full explanation from Barry, Kellan, and Krishna. Because a long-time partner was being acquired, we had 3 months to change the front and back end. I'm proud to say we successfully swapped out the engine and changed the wings mid-flight without losing altitude. In this open and collaborative environment, the team felt respected and valued, despite being asked to "give her all she's got." But unlike Scotty on Star Trek, we don't pad our estimates to look like miracle workers. We just treat each other, and our customers, with the respect and care we all deserve.

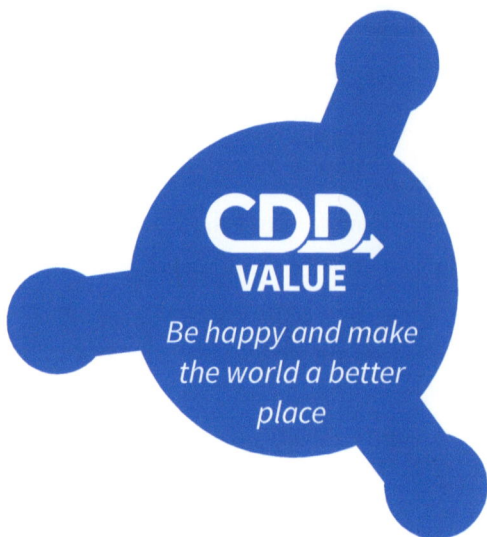

CDD.
VALUE

Be happy and make the world a better place

CHAPTER 16

Scientist—Jonathan Bisson, PhD

FOUNDER'S NOTE:

One of the many things I like about Jonathan is that he is a pure scientist and a pure programmer. This is rare. He is a strategic thinker. And such a warm and caring person. You may recall that I refer to Dr. Sylvia Ernst as the Oracle. Seeing there was something special in Jonathan and then going the extra mile to successfully recruit him is a credit to Sylvia for seeing the possibilities.

I also really like that he worked with Scott Franzblau on Tuberculosis (TB) drug discovery. TB has been one of the worst and most deadly diseases for humanity for centuries. At CDD, in addition to our program making the software available for scientists in Africa, we have been actively supporting the Gates Foundation on the Tuberculosis Drug Accelerator (TBDA) consortium bringing together nine international major pharmaceutical companies with leading government, academic and non-profit researchers.

⚛

I t all started on a stormy morning ... well, not really. The truth is, what led me to CDD wasn't a single event on a single day. It was a logical landing spot after traveling the world and switching career lanes throughout my life.

I'm much more of a generalist than a specialist. I like a lot of things, and my brain likes to chase the next new shiny thing. My work at CDD reflects that curiosity. I am a software engineer (a fancy name for a programmer), but with a focus on biology-chemistry interactions. And I never would have joined CDD if not for plants.

Two of my life's passions began in my early childhood: computers and plants. Around 4 or 5, I remember watching my Dad make our MSX computer screen change colors. This absolutely fascinated me. I wanted to be able to do that myself, but my parents were clear: "First you have to learn how to read." Well, if that's all it takes. I learned how to read and spent my time trying things, breaking things, dismantling them—and not just computers. I got shocked by 220v a couple of times, including the time I plugged a car alternator into the mains. If you don't know, plugging a standard car alternator directly into household mains will both electrocute and burn you—it is not designed for the high AC voltage from the electrical grid. I don't remember it, but my whole family does, as it hurt me quite badly. Around that same age, I learned that plants could be used to treat and heal the body and was proudly sharing around that I could use any plant from the garden. My parents replied, sometimes, but not all plants, so PLEASE DON'T TRY IT!

At school, I got an Advanced Technician Certificate in electrical engineering and worked briefly in the auto industry but rediscovered my love of plants and contacted various ethnobotanists for advice. The consensus was to "get a degree in medicine, biology, and/or chemistry." So I got a master's degree in biochemistry. I learned optics, physics, and quantum mechanics starting with the Schrödinger equation, and so on. This proved invaluable, and I liked the blend of technical work, research, and everything in relation to living things (or parts thereof).

I worked in medical device production and later joined the Institut de Chimie des Substances Naturelles (CNRS) in Gif-sur-Yvette near Paris. It was THE phytochemistry lab in France, working with plants from abroad and even collecting them themselves. It was here that I grew a little frustrated by the lack of tools to organize lab data. So I started creating my own tools to manage samples and experiments.

Meanwhile, my now-wife was waiting for me in Bordeaux, and I decided to apply for a PhD there to work on the phytochemistry of Vitis vinifera, the plant we make wine from and especially its relation to health. Later, I did a postdoctoral appointment in wine, separating it, tasting fractions, and searching for bitter compounds. This was fascinating, adding the sensory aspect instead of just numbers on a screen. Following that, I joined the University of Illinois at Chicago with Professor Pauli, director of the Pharmacognosy Institute and Institute for Tuberculosis Research. In my 9 years there, I worked on numerous projects from Dentistry to Parasites. I began to rewrite and manage NAPRALERT, the largest database of natural products and their traditional and medical uses. Together with Pierre-Marie Allard and Adriano Rutz, we launched the LOTUS project, which consisted of documenting everything that was known about which natural product can be found in which organism within the public domain inside Wikidata (the data pendant of Wikipedia). I also joined the Institute for Tuberculosis Research part-time, where I worked on cyclic peptides and mass spectrometry and started to get more involved in the data and sample management of the institute.

This is where I was first introduced to CDD by the director, Scott Franzblau. He was using it for TB Alliance projects and thought it could be a good fit for the lab. I reached out to CDD, we tried the CDD Vault for the lab, and I discovered that it had most of what I had been desperately trying to do by myself all these years.

Turns out data management wasn't an easy single-developer problem.

I helped our various users import legacy data and structure their experiments. Most importantly, CDD Vault reduced one of my frustrations during meetings: trying to find the data. In many meetings throughout my career, we spent 90% of the time trying to find who remembered where the data was for that experiment and how it compared to other data points. With a system like CDD Vault, it was a couple of clicks away and allowed us to focus on the real thing.

After nine years in Chicago, Sylvia Ernst reached out to see if I was interested in working at CDD. All the software seemed to work... so I wasn't sure what I would do there. I did the interviews anyway, and that's where I realized there were many projects in new areas of interest for me. More importantly, all the people I talked to were absolute experts and were also nice—focused on a collective endeavor. It may sound ridiculous, but from my meager experiences, this is not that common. I happily took the job.

Now, three years later, my wife and I are back in France, expecting our twins and still working with CDD. These are some of the most rare and precious things we got from our 12-year adventure in the US.

At CDD, I get to combine two of my lifelong passions by building software that helps researchers discover new medicines, many still derived from the natural compounds I have known and loved, while integrating state-of-the-art cheminformatics tools and AI models. It is deeply satisfying knowing that the platform I develop helps scientists organize, analyze, and share their drug discovery data, especially for the neglected diseases that continue to afflict the least lucky of us.

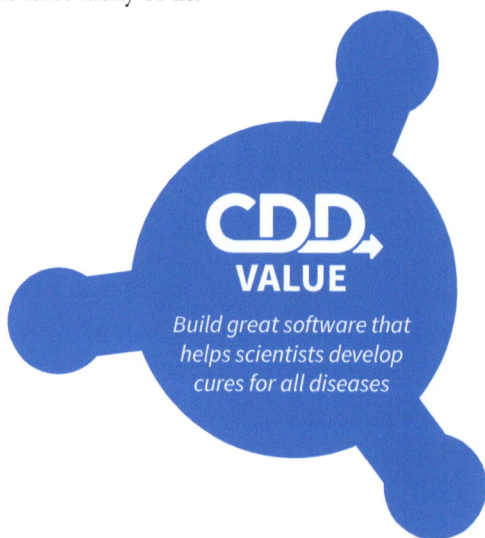

CDD.
VALUE

Build great software that
helps scientists develop
cures for all diseases

CHAPTER 17
Developer—Peter Gedeck, PhD

FOUNDER'S NOTE:

Usually, people are either experts in scientific discovery (0 → 1) or engineering (1 → 1000). Somehow Peter is expert at both.

He uses knowledge gained from many years at Novartis together within CDD's software environment to invent new-to-the-world capabilities that are intuitive for scientists to use.

The latest collaborative innovation is the Inference Model—zero click, continuously optimized, automated predictions. Often there is a rift between computational scientists and experimentalists. With these zero-click models researchers can mathematically see and assess predictions side-by-side with the experimental results. It no longer becomes a religious question of who is right (the computational scientist or the experimentist), because in this collaborative environment the scope and limitations of each model and experiment become self-evident.

The CDD team trusts Peter Gedeck's intuition, and the scientific community benefits. Via the millions of logins by the scientific community, those benefits broadly translate to global human health physically manifested in new medicines—from ideas to bits to atoms.

From an early age, I knew I wanted to be a scientist. With a strong interest in science and mathematics throughout my childhood it felt natural to pursue chemistry first at the University of Erlangen-Nürnberg in Germany and later at the University College London in the United Kingdom. It was during this time that I found a way to merge my interests through theoretical and computational chemistry.

My PhD research focused on photophysical processes that drive charge separation, an area directly relevant to solar energy conversion, which is currently relevant given energy needs and environment considerations. That experience deepened my appreciation for science that combines fundamental theory and real-world application. During my postdoctoral work, I shifted toward computational methods in drug discovery. This area combined everything: my love for science and mathematics as well as being able to contribute to a field of relevance.

When I left academia, I started working in drug discovery at Novartis. Even though the work was guided by commercial interest, pharmaceutical research is fundamental research and there were plenty of opportunities to develop and publish basic science. In total, I worked at Novartis for twenty years in the United Kingdom, Switzerland, and in Singapore. When I moved to the USA, I left Novartis and eventually joined CDD's research informatics group.

CDD is a great fit. In research informatics, we develop and explore cutting-edge technology. Although most projects are integrated into Vault, two projects developed into standalone products: Bioassay Annotator, an application for ontology-based annotation of biological assays, and PharmaKB, a knowledge base around pharmaceutical drugs. Let me point out a few of particular interest given my experience working in Novartis project teams.

In the intelligent chemistry browser inside CDD Vault, we developed a method to organize SAR (structure activity relationship) datasets using a fragmentation based structural hierarchy to see trends. We incorporated this technology into our free Visualization application in a way that does not expose the complexity of the hierarchy. The fragments of a structure are shown and can be used to quickly see non-obvious trends—like additivity and non-additivity underpinning molecular bioactivity patterns.

We also worked on another interesting project focused on the development of a deep learning network for the inverse QSAR (quantitative structure activity relationship) problem. In QSAR, we express chemical structures using numerical descriptors and use them in regression and classification models to predict desired properties. Using a model, we can change the numerical representation, a vector of numbers, of a given structure in a direction that improves the calculated property. Once there are known vectors, we need to recreate molecules. This is the inverse QSAR problem. Our solution was coupling a graph convolutional network as a numerical representation. And the reverse, a recurrent network that creates a structure representation from the numerical representation. This worked well but not reliably enough to expose to our customers, which is a much higher bar (100% fidelity requirement). However, we realized that the numerical representation has characteristics that make it optimal for ultrafast similarity-based applications.

This led to the development of the AI module, with two main features. The first one uses numerical representation as the basis for very fast similarity searches against large compound collections without exposing structural information to 3rd party websites. This allows scientists to learn more about their compounds, explore the patent (IP) space, and identify similar commercially available compounds. The second feature suggests variations of a given structure. Structures with similar bioactivity are referred to as bioisosteres. These bioisosteric suggestions are close analogues with relevant and chemically reasonable modifications, because it was trained on (and constrained by) real, existing molecules. This is in stark contrast to purely generative models, where the produced structures often have no relationship to the starting point—and can suggest nonsense molecules. The Generative Bioisosteres are generally ones that actually can be made.

While the technology itself is often abstract and its usefulness in drug discovery at first glance not obvious, the creative process is guided by the principle to achieve a product that will be used by scientists in an intuitive way where the complexity is simplified but not dumbed down.

Bringing a project from idea to product requires a variety of skills. It starts with the creative step that involves one or two people. But this only gets you to a certain point. Developing a product that is efficient, robust, and safe requires a highly functional team with broad experience working well together. We have that at CDD.

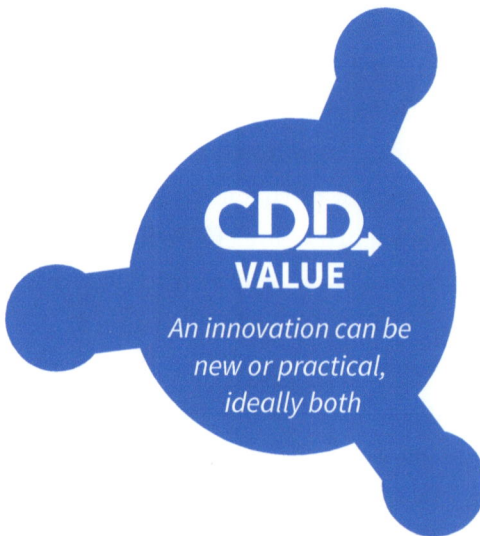

CDD.
VALUE

*An innovation can be
new or practical,
ideally both*

CHAPTER 18

Gamer—Jason Bret Harris, PhD

FOUNDER'S NOTE:

I had to let Dr. Jason Harris go when times were tight. At CDD we aim to be a kinder, gentler company. We kept the relationship, brought him back, now he collaborates with everyone and is a key piece in the puzzle. He has contributed everywhere from data analytics to research grants to website development to product management to accounting analytics to business development and more. I tend to get interested in lots of areas and need people like Dr. Jason Harris to follow through precisely on all the details. I'm glad CDD is here for Jason and that Jason is here for CDD. Everyone enjoys working with Jason; he is a good person.

CDD is an interesting place to work because we solve a lot of problems. This is the hallmark of any innovative and successful growing company like CDD. Fortunately for me problem-solving has always been at the heart of what I enjoy most.

They say: "do what you love, and you'll never work a day in your life."

I've spent more hours immersed in games than I'd care to admit, and for me, problem-solving, even at work, feels like just another game. What captivates me isn't simply the goal of winning, but uncovering the logic behind the puzzles, the hidden systems running just out of sight, and the rules that make everything tick. Games have always fueled my curiosity and sharpened my need for clear, logical conclusions.

Growing up during the emergence of home computing, with all its wonders and glitches, extended that enjoyment of problem-solving into screens and computers. Limited to a single Pentium computer and a stack of AOL trial disks, I have fond memories of sitting on the floor, looking up at my older brother, and waiting for a turn at the keyboard. He would usually hand off repetitive tasks, like collecting resources. We didn't realize it back then, but those chores sparked our shared interest in learning to code. Nowadays when we play computer games together, in addition to discussing rules and strategy, part of the fun is figuring out technical solutions to make the work more efficient.

In academics, I developed a strong interest in science. It offered a different kind of puzzle: the challenge of understanding how the natural world works. As a youth, I gravitated toward the natural sciences, since that curriculum was most accessible to me, but I never lost my interest in games and computers. These interests eventually came together during my PhD, where I built computational models to explore scientific problems.

I immersed myself in the logic of molecular systems, creating physics simulations to visualize biochemical interactions. The same curiosity and reasoning that made games so compelling could also be applied to unraveling scientific questions. Coursework, experiments, and models became quests, and each milestone felt like unlocking a new level. This dual love of science and technology ultimately shaped the structure of my thesis, where lab experiments were paired with long nights building models and coding.

That was where I first ran into the limits of technology. Models could do a lot but scaling them required something that was missing: data standards robust enough to let scientists combine, extend, and share their work. That realization shaped my postdoctoral research, where I shifted focus to a side quest designing and implementing data standards for more complex, powerful models.

That led me to CDD. Here was a place where data standards weren't a side quest, they were central to making science more collaborative. CDD is also a rapidly growing company, with new business challenges emerging all the time. Joining CDD gave me the chance to bring both sides of my character to the table: the scientist's eye for what research needs and the technologist's focus on efficiency.

While at CDD, I contribute to emerging technologies and drive the company's operations. One morning I might be creating data standards with experimental biologists, and by the afternoon I'm troubleshooting a patent application while also on another monitor in a Zoom call reviewing creative images. Later, on the same day, I'm helping accounting build a better management process. Describing all the games (err rather roles) I get to play while at CDD would be a laundry list . . . hopefully, you get the idea.

My career wasn't designed; it unfolded, one problem at a time. I turned play into purpose. My own journey moved from games to computers, from science to modeling, coding, standards, and ultimately from all of these into my various roles at CDD (informatics research scientist, grants compliance, digital marketing, website administrator, business intelligence analyst, customer resource management (CRM) administration, product manager, head of business operations). I didn't follow a plan; I followed my curiosity. It is proof that pursuing one's natural passions can grow into something that helps a company like CDD, and the world around it.

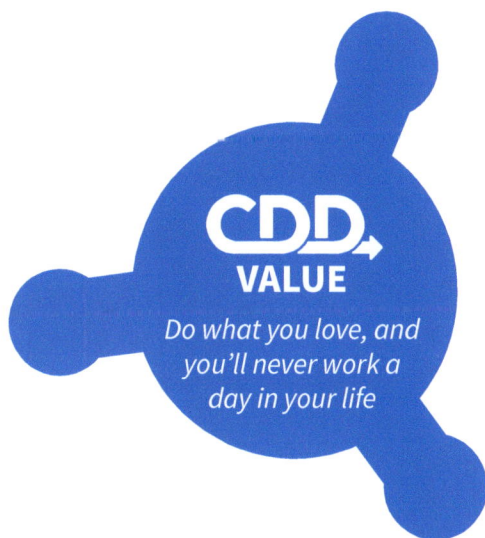

CDD.
VALUE

Do what you love, and you'll never work a day in your life

CHAPTER 19
Researcher—Alex Clark, PhD

FOUNDER'S NOTE:

Alex is a pragmatic purist. The fact that he thrives in a culture as collaborative as CDD is a credit to his own co-evolution with the team.

Most great engineers classify themselves as pessimists (finding bugs, mitigating risks, etc). Somehow Alex is an optimistic engineer. Like many others at CDD he is both a scientist and a technologist. He likes to do things rigorously correctly, but in a new, different, better way. This has led to breakthroughs in how we handle assays, mixtures, and "chemically aware" biologics.

B ack in the 1980s when I was ten years old I borrowed a ZX81 computer, plugged it into the family television, and typed in my first computer program.

I was completely hooked.

At university, I decided to study something I couldn't learn at home, which is how I ended up with a PhD in Chemistry. As much as I loved science I found that ultimately I'm more of a builder than a discoverer. Combining software engineering and chemistry into the same day job was not straightforward in the late 1990s and was made even more difficult by being in paradisiacal but remote New Zealand. Fortunately, the drug discovery industry has a strong demand for that skill combination, and I've been here ever since.

One of the greatest things about working with the engineers at CDD is that the group functions as a true team rather than in name only. The environment fosters individuality by enforcing rigor and discipline in all that we do. The success and camaraderie free us up enough to do more exciting work, and because we are encouraged to get involved in publishing and training, we get credit for what we've done.

Regarding the impact of our work, I don't personally believe companies have a moral "responsibility" to have a charitable component. Mission-driven small-to-medium sized companies like CDD exist to make things fundamentally better, which benefits everyone. Our responsibility is to succeed at what we do best and not get too distracted. If we do that, everyone wins.

I hope that I have and continue to contribute a sense of craftsmanship to our cottage industry, which is niche but important. Scientific software should solve new and original problems while holding up aesthetic ideals, e.g., being beautiful to look at, comfortable to use, familiar to learn, elegant to maintain. These goals will always be aspirational but every so often there is a product or a feature that really stands out.

In engineering, our mission is precise: build something new—and do it right.

I crossed paths with CDD near the end of its first decade. My involvement with the company could best be described as orbital capture. We worked together for a while, then some more, and then a lot. Before too long it was all the time. Over the course of CDD's ensuing decade there are three projects that highlight our evolution from a lean and dedicated startup to a recognized industry leader.

Without going into all the details, which you can read in our publications, blogs and conference presentations, these three projects provide superior data representation of bioassays, chemical mixtures, and "chemically aware" biologics.

The first project for bioassays is an interesting two-way capability. One can take human readable text describing a bioassay and convert it into a meaningful, computer readable data format. And one can also do the reverse, take a list of attributes and generate readable text. For context, this all pre-dates ChatGPT. Why is this important? Because when scientific knowledge is encoded into a data structure, it can be truly used by computers, rather than just stashed as an archive. For example, in bioinformatics, one can look at similar sequences of nucleotides or amino acids for similar properties. In cheminformatics, one can look at similar molecules for similar properties (potency, selectivity, toxicity, etc). Now for the first time, researchers via assay informatics can look at similar assays. They can now compare apples to apples.

The second project—chemical mixtures—was going from 0 to 1. There was no general format for handling mixtures, despite the fact that the overwhelming majority of chemicals are encountered in mixture form in almost every context. Now there is a format, and it is clear, open and standardized. This has applications from drug screening (think combination therapies) to formulations (think pills) to consumer products (think toothpaste).

The third project was super exciting, since we could precisely handle small molecule mixtures, we realized many biologics are mixtures, too. Natural antibodies in everyone's immune system are a combination of two light chains and two heavy chains—now one can represent all four, by their sequences (bioinformatics), their atoms (cheminformatics) and the connecting disulfide bridges that hold them together. So "chemically aware" biologics are interesting for natural biologics and for any modified biological by humans beyond what occurred during the billions of years of evolution. The real breakthrough is atomicly precisely representing every variation achievable with human (or machine) intelligence atop of nature's evolution. For example, a hot area for cancer therapies is Antibody Drug Conjugates (ADCs). Here one can generate huge antibody libraries to gain specificity for binding only to cancer cells (and not healthy cells) with a toxic small molecule covalently attached to the antibody to precisely kill those cancer cells. The amazing thing with "chemical aware" biologics is now users can optimize all components—the antibody, the linker, and the toxic molecule (referred to as a warhead). Because every atom is precisely represented in this single combination therapy, they have the form they need—whether they want to look at biology, biochemistry, or chemistry.

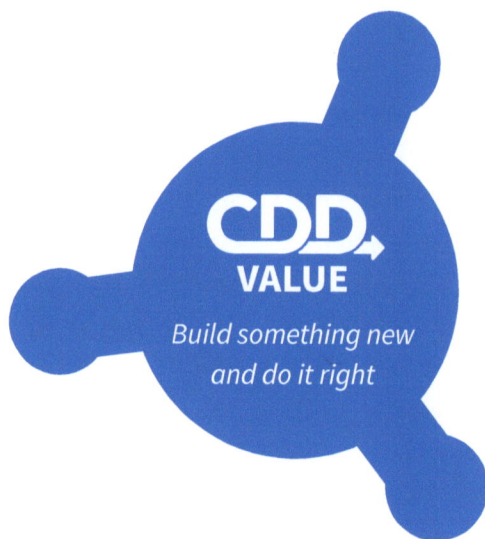

CDD.
VALUE

Build something new
and do it right

PART 6

COLLABORATORS

CHAPTER 20
Artist—Joe Clifford, MFA

FOUNDERS NOTE:

Joe is one of my favorite undiscovered artists. I named my dog Lucky, after a song where Joe sings, "How Lucky will be the only one who misses him when he's gone." He pours his heart into his music and books and paintings.

I knew Joe first and foremost as a musician. My next-door neighbor growing up was his bass player. It was obvious Joe had talent and somehow also had unique challenges to overcome. He transitioned from being a professional musician to a writer, with his breakout book, "Junky Love." He has now written over 20 books, anthologies and short stories—the creativity never stops. I like his books, but I love his music. In his most recent song "A Better Machine" you can hear his raw feelings when he sings, "I see the kids playing on the playground, and I wonder what this world could have done. I see the way my boys look at me now, like I'm good enough, this thing I've become." This book wouldn't have happened without Joe, and for that I'm forever grateful. But really it is much more. Great artists and scientists try to see the world as it really is. And truth can be beautiful and painful. "Cry me a River" and "Are you Happy Now?" are perfect titles of his more vulnerable songs that never made it to a record but should have. Maybe he'll record them with his new band. I'd buy it.

The cornerstone of collaborative drug discovery is collaboration—which in our case are customers, vendors, and... collaborators. A great collaborator is one that goes beyond the face value, beyond the money side which measures and pays for value. We think internally and externally about the multiplier effect. The intangible positives of the interaction that flows to others for maximum positive impact. In thinking about external collaborators, I thought deeply about the people who do more than influence the direct people they touch. Through their words and actions, they touch many more... collaborators. The multiplier effect. We are honored that Pat Walters, Chris Waller, and Joe Clifford provided their personal backstories in the chapters that follow.

In the beginning, there was Barry.

I'd moved from my small farm town of Berlin, Connecticut, setting out for the big city of San Francisco. The year was 1992. The reason for my choice of destination was two-fold: the first was Jack Kerouac, for whom my second son (Jackson Kerouac) is named. The other is a lyric from The Replacements, the last year they were unassailably cool. "Left of the Dial," off Tim.

Headed out to San Francisco /
Definitely not L.A.

At twenty-one, I'd dropped out of college and settled on a profession. Rock star. The hours and lifestyle fit my artistic temperament. (I'd never been big on 9-to-5 work.) I'd fronted a band back east for several, semi-successful years. No, Redheaded Stepchild wasn't a household name. We had been in the Life & Leisure section of the *Hartford Courant*, our music played on local (left of the dial) radio stations, and we had a big gig lined up at the legendary NYC punk venue CBGB's. (I left before we played that show.) I really felt our songs, which I mostly wrote, were ready for the big stage and bright lights.

Of course I was wrong. (More on that shortly.)

Soon as I hit the city, I ditched the girl I'd moved here with and set out to find a new band. Back then, I didn't know I wasn't special. In fact, it was just the opposite. I believed I was destined for great things. I had no proof, no evidence, other than my mom convincing me I was meant to save the world. Which is another song lyric. But most of life comprises song lyrics, pop culture nuggets, gems culled from the peripheral, these pearls of profoundly obvious wisdom snatched up out the ether, what I used to call "yellow brick cracks." Not sure what I ever meant by that. But back then I treated my entire life like a blank canvas, which is the kind of pretentious crap you can get away with saying at twenty-three.

That's how old I was when I started my San Francisco band The Creeping Charlies.

I'd come in at the tail end of a rather monumental rock-and-roll event in San Francisco. I wasn't aware just how monumental, as self-absorbed as I was in those days. I could miss little details like replacing a singer who was on the verge of fronting perhaps the biggest band in the world (the Counting Crows). I didn't know Adam Duritz, had never met him, didn't know when I answered the ad in the SF Weekly looking for a singer that I'd be hooking up with the drummer and guitarist who'd written "Round Here" with Adam. I'm not sure that bit of trivia fully registered for years, if I'm being entirely truthful.

When I left Connecticut, I was also abandoning the cloistering Christianity I'd been raised on, exchanging it for head-on, full-tilt rock and roll rebellion. The drugs and waywardness were not far behind. The drummer got fed up fast, but the guitarist, Dan, stuck by my side. New drummers would come and go like Spinal Tap. But the big addition was our new bassist, Matt. Matt turned out to be lifelong friends with one Barry Bunin.

The Creeping Charlies never made it big. In fact, it can't be said we even garnered a small, cult-like audience. We did have one very loyal fan.

Barry.

He was at every show (most of which were held at the now defunct Hanno's, a hole-in-wall bar in an alleyway in the Mission). Barry was always front and center. Barry knew every word.

And later, after I'd return from the dead to restart the next incarnation, Barry would be there, too.

Being an artist, I can't tell you what it means to have someone believe in your vision. A good chunk of the artistic process involves sensitivity. A writer, musician, painter, whatever requires sensitivity to make the requisite connections with the world at large. If an artist can't find a way to make their work resonate with others, what's the point? Art is communication; it cannot occur in a vacuum. Of course, with sensitivity comes heartbreak, especially in the arts, which are rife with rejection. Barry supported my music, my painting, and my writing. And sometimes when you're slow to find faith in yourself, that kind of belief is the one thing to keep you going.

When Barry floated an idea about Collaborative Drug Discovery and asked if I'd be interested, I was intrigued if wary. My relationship with corporate America, which for our purposes here we can define as pretty much any traditional nine-to-five, has always been prickly. We artists don't like clocks. We don't work well within the confines or parameters that govern employment. Calendars go unused.

My main gripe with the corporate world is that I see a wide divide between corporate interests, which revolve around the bottom-line, and any charitable component—what we owe our fellow man (in the universal sense). That's what ultimately piqued my interest in this book and CDD. The more I spoke with contributors (and I think that word works better here than "employees" or "workers"), the more I saw Barry had created something unique, instilling a pervasive value, shared by all, that profit doesn't need to come at the expense of personhood. In short, CDD highly values the notion of "giving back." There is an underlying precept that runs throughout CDD honoring this responsibility. You read it in these stories; you come across it in the underlying themes.

Of course a business, which CDD is, needs to be profitable. I'm not championing hippie ideology (I've never been much of a Grateful Dead fan). Part of why I became a mystery writer is that I like to make money on my books. (Genre fiction is more lucrative than literary navel-gazing). I don't see "commercialism" as a dirty word. But paramount: I demand my work speaks to something greater. I aim to tell the stories of the lost and hurt, the destitute and wretched, those wounded souls I met while living on the streets. My goal is to report back on what that life is really like, and, no, it's not all bad. I met some addicts who loved their children too. Not excusing use around kids, obviously. I'm simply saying that when you get to know people on their level, see them for the struggles they endure—when you watch folks who have less doing the most with what they got—it gets harder to be judgmental and outright dismissive.

That's my mission statement as an artist and writer.

Barry and the incredible talent he's assembled at Collaborative Drug Discovery operate in a different world, yes. But it's this shared ideology of doing what we can to help others—as much as we can, when we can—even if it might mean fewer bucks in the bank account—that I want to celebrate.

Maybe I am just an old hippie. But I'll go on record (and to my grave) believing that we've been put here on this Earth, graced with the gift of life (that at times can feel like a burden), to help one another make sense of the journey best we can.

And I feel confident saying, with that being the case, I fit right in with the folks at CDD.

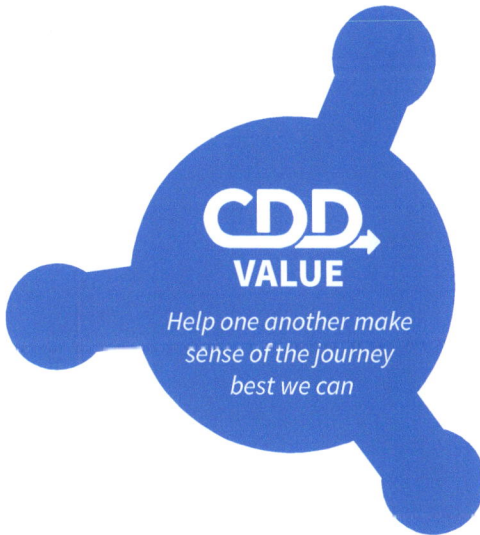

CDD.
VALUE

Help one another make sense of the journey best we can

CHAPTER 21
Ketcher—Chris Waller, PhD

Dr. Chris Waller is one of the funniest, most thoughtful, and creative leaders in our industry. Chris is also one of the rare people whom I've been across both sides of the table—both buying from and selling to. Although business serious conversations throughout, I always felt that I grew personally from the interactions.

When I was selling to Chris, at Pfizer, where he had great responsibilities, we started with science before business. We together had demonstrated on the world's largest set of ADMET (Absorption, Distribution, Metabolism, Excretion, and Toxicity) data run under standardized conditions that open source descriptors and models performed statistically equivalent to expensive commercial ones (similar kappa values). This was fascinating, and not something trivial to get through the delicate legal review for a press release with a smaller company. So instead, we co-published it in a peer reviewed scientific journal, with Pfizer and CDD scientists as co-authors (including Dr. Eric Gifford who traveled with Chris to other companies and later to CDD). Of course the structures were hidden, but the message was clear: everyone was bragging about their models being the best, yet all the models performed similarly. The emperor had no clothes.

When buying from Chris, it was also a collaboration. Expanding the pie, as they say. While CDD was building the largest, profitable commercial hosted private drug discovery platform in the world, together with Chris' team at EPAM we were co-creating the most robust, open-source front end for drawing small molecules which we have now extended to "Chemically Aware Biologicals." This is a major advance of the state-of-the art, unlike in the past when biologics were simply represented by their alphabet (GCATGCAT, etc. for oligonucle-otides, Ala-Val-Thr-Pro, etc. for proteins), now for the first time researchers (and programs) can store, search, and optimize based on every atom in every large biological—natural or hybrid.

In both cases the public is richer, and so is my heart.

In 1987, I graduated from Davidson College with a degree in Pre-Medicine and absolutely no desire to go to medical school. So, I became a parents' worst nightmare and moved back home. I spent 9 months working in the Quality Control lab in a process chemistry company working 12-hour swing shifts. Medical school was looking better with each passing hour, yet I didn't want to practice medicine. I wanted to do medical research! So, I chose UNC-Chapel Hill, and earned my doctorate degree in Medicinal Chemistry, followed by a post-doctoral fellowship at Washington University in St Louis, where I focused specifically on methodologies for the design/prediction of bioactive compounds. At the end of my fellowship, I had another deci-sion to make—industry, academia, or government? I chose the government, to "return" to the EPA to finish work that I started there in a part-time job during my graduate career.

Later, in the mid-90s, the biotech revolution was in full bloom, and I was anxious to get involved, so I joined a small, NY-based firm known for oncol-ogy research. The company was attempting to make the move from a contract screening company to a fully integrated drug discovery company (FIDDCO). After three years, our team delivered the first integrated informatics platform supporting all aspects of the early discovery process, the first commercially available HTS robot (predecessors of the uHTS robots in commercial use today), all integrated with a compound library design and synthesis platform.

When you leave a job, and you will, do it as gracefully as possible. The world is very small—you will see these people again. After 3 years of flying up and down the east coast and back and forth to England, I was tired. I decided to step away from the "port-swilling, cigar-smoking, deal-making" life and took a position at Sphinx (Lilly), focusing specifically on the design and implementation of informatics systems to support the lead generation chemistry group in North Carolina. But in the aftermath of 9/11, travel became very difficult. With much of my family in Michigan, I left Lilly for Parke-Davis (Pfizer) in Ann Arbor.

I spent 11 years there, starting in the birthplace of Lipitor doing exactly what I thought I was trained to do. I had a Director-level position in a discovery research organization with no direct reports but a large team of very talented computational chemists and informatics professionals at my disposal. And. . . a seemingly unlimited budget. I spent four years as a chemistry informatics (IT) leader. I was once again blessed with a team of highly talented, diverse, and driven individuals and we changed the world—OK, Pfizer chemistry. I convinced my management that I could change the world outside of Pfizer by promoting the Pistoia Alliance—if only I was given the chance. I was given the opportunity with a one-year leash and no ability to return to my previous role. A classic hero or goat situation. Halfway through this, Pfizer imploded and management shake-ups were happening. My manager called. "The bad news is that I've been fired. The good news is that you haven't—yet." At my first meeting with my new manager, the discussion went something like this. . .

> **Manager:** "Who are you and what do you do?"
> **Me:** "I change the world by convincing people who otherwise would not collaborate to do so by giving them a safe pre-competitive environment where they can share ideas."
> **Manager:** "That sounds like a great job—we don't have that here."

The next few years were spent in cycles of reinvention. Leaving Pfizer in 2012 was very liberating. I was free. Free to think. Free to explore. I started a private consultancy. Independent consultants need great networks, not just for clients, but also for collaborators and providers of technology (turns out that not all the software that cheminformaticians use is free). I began working with the Medicines for Malaria Venture on a cool project to build QSAR models using data from disparate sources that couldn't be easily combined. Barry and the folks at CDD were very helpful in getting me ready access to the data, and by then I was sold on the value of collaborative data sharing platforms. Honest and impartial data brokers are critical to the success of federated learning. That project followed me into my next career move.

I joined Merck & Company and found that the world is a very small place. My manager, Frank Brown, was an old friend and mentor from my UNC graduate school days. Over the next 5 years, my team delivered a fully functional cheminformatics platform for Merck, which formed the basis of predictive science capabilities spanning research, development, commercial and medical functional domains. Again, I was blessed to be surrounded by incredibly passionate professionals and together we changed the world— OK, Merck. I was able to recruit smarter scientists to work on (and complete) the work that I had started with the MMV. The CDD platform is now being used to collect and manage data created at Merck for various projects in the not-for-profit space.

In 2017, I joined EPAM—a company that I had worked with while at both Pfizer and Merck to build informatics platforms. I characterized (and justified) this move as an effort to learn how the sausage was made. I embarked on a one-year experiment that lasted a little over 8 years. I was able to continue working on pre-competitive projects, delivering open-source tools as accelerators of commercial and bespoke informatics platforms. This presented another opportunity to work with Barry and the CDD team, this time in a collaborative software development project that saw the incorporation of EPAM open-source components into the CDD Vault platform to enable, to date, the best solution for the management of multimodal therapeutic agents. Without going through all the details of my EPAM experience, I can now safely say that I am a master sausage seller with in-depth experience on making the tastiest sausage on the planet. Experiment over. You don't really want to know how sausage is made. It's very messy.

And while this book is going to press, I've joined Quantori!

Here's what I tell my children, grandchildren, team members, and pretty much anyone on the street who makes the mistake to ask me for advice. Do something meaningful—to you and to the world. It should be something that you like to do—you're going to spend a lot of time and energy doing it. Be the best that you can be—mediocre isn't good enough. And, very important, find somebody to pay you to do it.

Our industry is changing—driven by various pressures. We're behind, but we're learning. Advances in chemistry, biology, computing. . . are all changing the way we work, many of these changes are evolutionary (and almost predictable). It's the non-traditional technologies and trends that will revolutionize the industry.

We're learning from the new generation, depending on them to bring the softer skills related to interpersonal interactions—collaboration as the new normal, crowd sourcing, crowd funding, social media, and generative AI. I'm seeing exciting things happening across the research, development, commercial, and medical continuum—largely experimental, but the revolution is coming. Much of the future I see already happening at CDD—supporting secure collaboration globally, next generation drug discovery informatics capabilities for small molecules and biologics, and of course AI. I am honored to be one of CDD's long-term collaborators.

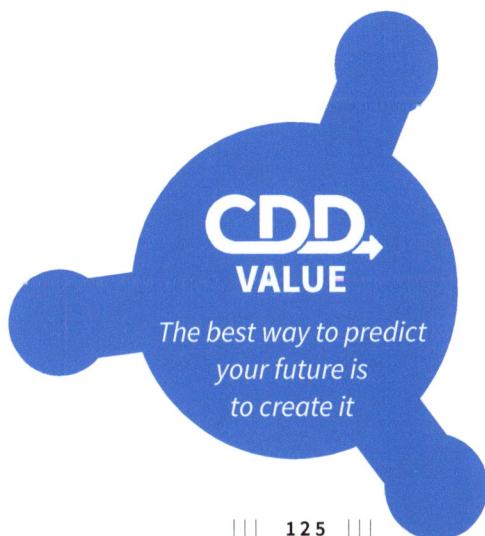

CDD.
VALUE

*The best way to predict
your future is
to create it*

CHAPTER 22
Babel—Pat Walters, PhD

FOUNDER'S NOTE:

I can't tell you the number of highly critical, perfectionist leaning scientists who are difficult to work with—it is a cliche. Like the actor who needs his own trailer prepared just so. On the other hand there are people so nice and kind, but they never get anything done. What's remarkable about Pat is that he is so thoughtfully critical, yet at the same time so universally adored. I think it is because all his lessons (and code in GitHub: https://patwalters.github.io/year-archive/) both speak truth to power while somehow being accessible to both the most advanced experts and enthusiastic beginners in the field. It takes a true master of one's art to speak to both, which is exactly what Pat is and does.

began my scientific career as an experimentalist in polymer chemistry. Although I enjoyed working in the lab, my first job after college involved a long sequence of tedious manual steps. Not being a patient person or someone who liked monotony, I decided to explore how I could use laboratory automation to streamline my workflow. It was the early 1980s, and since there wasn't software available to do what I needed, I chose to write it myself. I started programming in BASIC and quickly switched to Turbo Pascal. I was excited by how much this new skill enabled me to accomplish. I could write a program that connected autosamplers, switching valves, and HPLC systems to perform analyses overnight while I was sleeping.

My early programming experiences also taught me the value of open-source software. This was before the internet, but I could still join clubs and user groups where I could order floppy disks with code for useful routines. I aspired to write code someday that others would find helpful.

After a few months of programming, I became hooked. While continuing my work in the lab, I created a database to store the results of my experiments and implemented a network to connect the computers and instruments in the lab. This is routine today, but 40 years ago, it was revolutionary. I was eager to learn more about this computer that was transforming my work. I devoured computer magazines and visited the computer book section of my local Barnes and Noble several times a week. During one visit, I found a book with accompanying software titled "Turbo PROLOG—the natural language of artificial intelligence." Intrigued, I bought the book and software and started learning to program in PROLOG. Before long, I built an expert system to help the company's technical support team recommend analytical methods. Around the same time, the company I was working for was acquired by a larger organization. I was fortunate that this new company had an AI group I could work with and learn from.

After a few years, I decided to pursue programming as a full-time career. I loved chemistry and wanted to find a way to combine my two passions. After a chance conversation, I was introduced to Prof. Dan Dolata, who led the AI in the Chemistry Lab at the University of Arizona and became my graduate school advisor. While pursuing my PhD, I worked on various AI applications, ranging from algorithms for conformational search to machine learning projects. Once again, open-source software played a significant role in my career. Much of my early work in graduate school involved converting between file formats used in molecular modeling. To help with this, my labmate Matt Stahl and I developed a program capable of translating between some of these formats. Other students and postdocs in my department saw it and asked me to add more formats. Over time, the program became broadly useful, and I wondered if others would find it helpful.

This was the early days of the internet, and websites were just beginning to gain popularity. I posted a note on the Computational Chemistry List (CCL) to let people know that the program was available. I created a simple website to host the code and asked users to register before downloading. I was pleasantly surprised at how helpful people found the program, Babel, named after the Babel Fish in *Hitchhiker's Guide to the Galaxy*. Within the first month, we had more than 500 registered users, and over 1,000 by the end of the first year. Babel also opened doors for me; being known as the "Babel guy" helped me land several job interviews. One of these interviews led to a position at Vertex Pharmaceuticals, where I worked for the next twenty years.

In my early days at Vertex, I mainly focused on developing computational methods. This marked the start of virtual screening, and my colleagues and I developed techniques to identify drug-like molecules, along with initial applications of machine learning to properties like blood-brain barrier penetration. Mark Murcko, my boss and mentor, was excellent at guiding my efforts toward interesting and relevant problems. When I joined Vertex, the pharmaceutical industry was just beginning to explore the field of combinatorial chemistry. Companies like Chiron, Merck, and Alanex were developing methods to analyze and optimize combinatorial libraries. As we refined these methods, we needed reliable examples of combinatorial reactions for our analyses. At the time, the literature was scattered, with no standard reference for combinatorial chemistry. Enter Barry Bunin, who conducted pioneering work in the field as a graduate student in Jonathan Ellman's lab at UC Berkeley. Many in the field celebrated the publication of Barry's book, "The Combinatorial Index," which became the definitive resource for both computational and synthetic chemists.

In 2016, I left Vertex to become one of the first employees at Relay Therapeutics, a company focused on tightly integrating computation and experimentation. After leaving Vertex, many of my former colleagues who had moved on to other companies contacted me to ask about scripts I had written in the past. Since it seemed there was an audience for some of the cheminformatics and machine learning work I was doing, I decided it might be fun to write a blog. In 2018, I published my first blog post describing Free-Wilson analysis, a simple yet effective method for SAR analysis, originally published in 1964. To accompany the blog post, I created a GitHub repository with a code example. My blog, Practical Cheminformatics, gained some popularity and led to a series of interactive cheminformatics tutorials. One of the most satisfying aspects of writing the blog and tutorials is meeting students and early-career professionals who tell me how my writing has influenced their career paths.

In 2023, I received the Herman Skolnik Award for Chemical Information Science from the American Chemical Society, and my path again crossed with Barry Bunin. By this time, Barry's company Collaborative Drug Discovery (CDD) was well established as the industry leader in chemical and biological data management. I had long admired Barry's commitment to academic and open science and his support for open-source software tools like the RDKit. I was thrilled when Barry offered to speak at my award symposium at the American Chemical Society National Meeting.

Fast forward to today, and my journey has once again crossed paths with Barry and CDD. In 2025, I started a new role leading OpenADMET, an open science project that integrates insights from high-throughput experimentation, structural biology, and machine learning to enhance predictions of drug absorption, metabolism, excretion, and toxicity. All experimental data produced by OpenADMET is stored at CDD. Moreover, CDD publicly hosts the data, making it available for predictive challenges that allow scientists worldwide to test state-of-the-art machine learning models.

Spending the last 40 years working at the intersection of chemistry and computation has been an incredibly enjoyable and satisfying experience. Interacting with people like Barry and his team, who share this passion, makes it even better. I've been privileged to see the impact of computational methods, and now AI, on drug discovery. I've also seen the profound effect that new medicines can have on patients' lives. I'm a lucky guy.

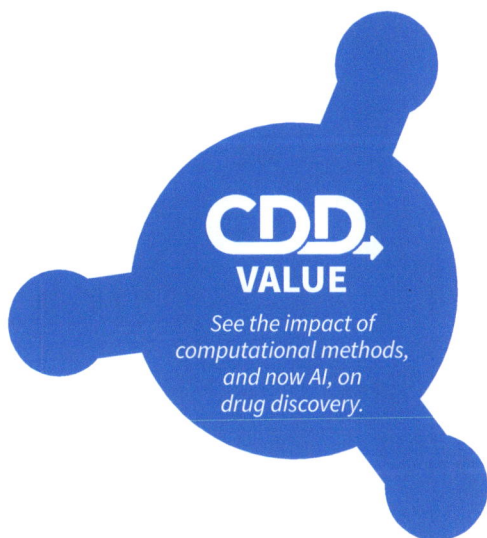

CDD.
VALUE

See the impact of computational methods, and now AI, on drug discovery.

EPILOGUE

Values

FOUNDER'S NOTE:

Abraham Wang wrote in his personal vignette a paragraph that represents how we've translated our idealistic vision into a profitable, sustainable business while staying true to our Values and Mission (the ideal of maximum good for maximum people):

> *"I am truly inspired by what the CDD team has achieved. Barry has found a way to preserve his original goal of doing good by humanity. Many founders set out to change the world only to get corrupted by the system. CDD has pioneered a business model that retains its independence and its culture, while being commercially sustainable for twenty-plus years."*

There is a saying, "Culture eats Strategy for breakfast." It starts with having values and a mission which are inspiring. By definition, to be more effective than the competition, the group dynamic must be better than just throwing money at a problem. It was in India, where I saw poverty firsthand, people with real problems and I decided, no matter what, CDD would help both for the poor and the rich. We put it right in our mission statement and have found ways to support global collaborations around the world consistent with supporting both missions, in harmony. CDD's mission, methods, priorities, and values were of course created—collaboratively:

MISSION

Provide an unparalleled experience for humanitarian and commercial collaborative research and discovery.

How: Our people, our values, our goals, our innovation, our technology. At CDD, our greatest asset is our collaborative team. We're committed to customer success because when you succeed, the whole world benefits.

Priorities: CDD is a mission-driven business that puts customers first, employees second, and company third.

HONESTY
collaboration
innovation HUMILITY
passion meaningfulness
RESPECT professionalism
SUSTAINABILITY

VALUES
CDD Corporate Values

COLLABORATION

We thoughtfully and strategically collaborate with all, both internally and externally.

INNOVATION

We invest our thinking and resources into change that will propel the industry and our company forward.

MEANINGFULNESS

We positively improve the process of collaborative research by offering our innovative tools.

HUMILITY

We never think we have all the answers. We listen to our customers' needs.

RESPECT

We treat customers, partners, providers, and coworkers as we want to be treated.

HONESTY

We keep to our corporate commitments and promises. Always striving to do better, we admit our mistakes and collaboratively build a solution.

PROFESSIONALISM

We are thoughtful, clear, and effective with internal and external communications.

PASSIONATE

We take pride in what we do.

SUSTAINABILITY

We are a business that must earn money in order to continue to support our customers and provide a place for our employees to thrive for the long term.

KINDNESS

Be kind. Kindness builds trust, relationships, and long-term opportunities.

VISION

We are a company that continually refines our vision of the next value inflection events at the organizational, departmental, and individual level impacting customers, employees, and CDD.

meritocratic
FUN Inspiring
motivational agile
empowering FRUGAL
CHALLENGING respectful
NURTURING
rewarding

CDD Workplace Values

All CDD employees strive to create a work environment that is:

FUN

We do not take ourselves too seriously. We celebrate our successes together, big and small (5pm board—woohoo!).

MOTIVATIONAL

We are empowered to develop our skills and talents, and are eager to apply them to do more and improve what we've started.

INSPIRING

We feel inspired to ideate, iterate, communicate and grow.

NURTURING

We cultivate talent, purpose and growth. We support each other in this process.

CHALLENGING

We push ourselves and others to become better and achieve more. As individuals and a team, we are part of a continuum, striving to improve ourselves, our products, and our processes.

EMPOWERING

We maintain open communication and let the information flow, as appropriate, so that others can help us or support us. We do not point fingers but instead feed forward. We embrace mistakes as part of learning, to become better persons and a better organization.

MERITOCRATIC

We base our decisions on evidence. It is not important who got the idea, only that it is the best. We take the best ideas and iterate on them to make them even better. We strive to display optimal judgment on decision making, while being especially open-minded to new experiments (mutations that help us evolve).

AGILE

We are able to change directions quickly to adapt to the market and its needs. We communicate tirelessly so that everyone is in synergy. We let ideas develop within our resource constraints.

RESPECTFUL

We respect and embrace each other's differences. Our diversity is our strength.

FRUGAL

We spend wisely and observe budgets. We provide the resources to get the job done.

REWARDING

We recognize emotional, personal, and professional success as an important reward. Financial success is not our only metric, but is proportional to our achievements.

A SHORT HISTORY OF CDD

CDD Vault is a hosted database solution for secure management and sharing of biological and chemical data. It has become the world's most widely used platform for web-based collaborative drug discovery with over 10 million logins and more than 4 billion experimental data points. A short history of two decades of work can be seen in the "CDD timeline" below:

CDD Milestones

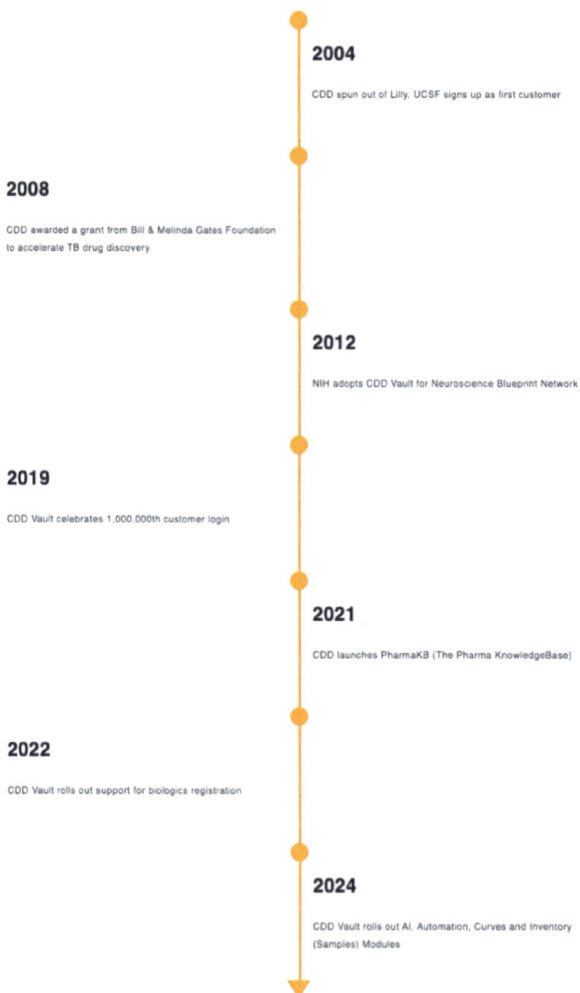

2004

CDD spun out of Lilly. UCSF signs up as first customer

2008

CDD awarded a grant from Bill & Melinda Gates Foundation to accelerate TB drug discovery

2012

NIH adopts CDD Vault for Neuroscience Blueprint Network

2019

CDD Vault celebrates 1,000,000th customer login

2021

CDD launches PharmaKB (The Pharma KnowledgeBase)

2022

CDD Vault rolls out support for biologics registration

2024

CDD Vault rolls out AI, Automation, Curves and Inventory (Samples) Modules

Cumulative Logins

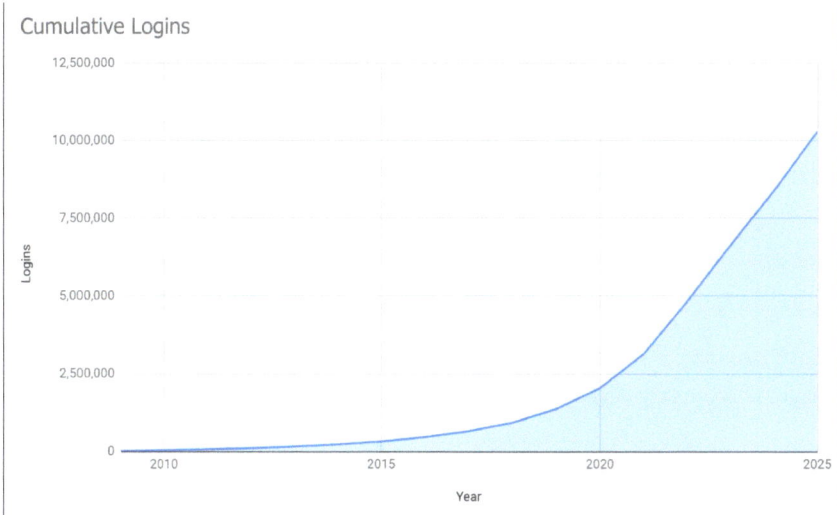

CDD Vault has witnessed exponential adoption over the last decade. The proof that a technology is useful is that people are using it. And it is worth mention- ing that this community adoption has happened with 100x capital efficiency (<1/100 the capital of the closest competitor). The business model works. One does not need to compromise values to win!

Nvidia CEO Jensen Huang with CDD Business card in his left hand at the JP Morgan fireside chat, onstage with Stripe CEO Patrick Collison. The CDD case studies with NVIDIA are 50/50 for humanitarian and commercial use cases, following a press release on the same day competitor Dotmatics (a combination of a dozen leading companies for drug discovery informatics) sold for $5.1B to Siemens.

CDD.VAULT®
Complexity Simplified

NVIDIA.

"The bioisosteres generation we can do using the CDD Vault AI Module is great for triggering new ideas and structural insights."
- Dr. Lori Ferrins, Associate Professor, Pharmaceutical Sciences, Lab for Neglected Tropical Disease Drug Discovery, Northeastern University

Customer Case Study

N Northeastern University

CDD.VAULT®
Complexity Simplified

NVIDIA.

"We use computational modelling to generate starting structures where they don't exist. Even when there are structures, we can use tools like AlphaFold to look for different forms and to carry different protein models forward."
- Josh Pottel, Ph.D., President and CEO of Molecular Forecaster Inc.

Customer Case Study

molecular forecaster

April 2, 2025

Collaborative Drug Discovery Integrates CDD Vault with NVIDIA BioNeMo NIM For AlphaFold2 and DiffDock Models

CDD NVIDIA joint case studies with humanitarian and commercial scientific collaborators.

CDD Spotlight Interview with our first customer—Dr. James McKerrow

FOUNDER'S NOTE:

In thinking about our values at CDD, and our ability to "walk the talk," I think back to the inspiring Professor-Doctor-Dean James McKerrow (PhD, then went back for his MD, most unusually). In addition to being our first customer, working on African Sleeping Sickness, Chagas Disease, etc, introducing us to the Gates Foundation (which has allowed us to support the double bottom line of diseases of the rich and poor)... I also think of him dancing at the Gordon Conference, playing in a band, and more. Jim McKerrow was especially unusual, in that he was willing to share his grant funding with others, while maintaining rigor in his science and passion helping on protease inhibitors (and other MOA) for NTD. We have dozens of CDD Spotlight interviews with many of our hundreds of customers. Jim McKerrow was appropriately the first one; he was the first one to take a chance on CDD.

"I THINK THE CHALLENGE NOW, WHICH YOU'VE HINTED AT, WILL BE TO BETTER COORDINATE THE EFFORTS OF THE GROUPS AROUND THE WORLD."

Interviewed by Barry Bunin, PhD, CEO, Collaborative Drug Discovery, Inc. on 29 March 2011.

Barry Bunin: My first question is just how someone who's an MD got interested and decided to do research because actually my family is all MDs. My dad's a family physician practicing doctor, my grandpa, all my uncles, and even my brother has a free clinic in LA and they're all seeing patients and you're doing research so...

Jim McKerrow: Yes.

Barry Bunin: I'm just curious how it came to be for you.

Jim McKerrow: Sure. Well we can start there and that's an easy answer because there's actually more to the story. Actually, I also have a PhD but I'm not an MD-PhD. I went to graduate school first in biochemistry at UCSD to get my PhD after college and my thesis project was on peptide chemistry, so I was really more of a chemist with a biological bent. Following graduate school, I had intended to do a postdoctoral fellowship in x-ray crystallography—but for a variety of reasons, mainly because I was mostly interested in applying biochemistry and structural biology technology to issues of human health and disease, I was talked into going to medical school to get a broader perspective on clinical issues and human disease. So I did the weird thing of instead of doing a postdoc, I went to medical school to get my MD. So I was not really an MD to begin with, I was actually a PhD scientist to begin with. And when I went to medical school, I had every intention of getting back to a lab as quickly as possible—but after I finished medical school, my medical school advisor said, "Well no, society has just invested all of this money in your medical education, you need to do postgraduate training in medicine," so I did an internship in internal medicine and I did a residency in pathology. I did some elective work during my residency in infectious disease, which is one of the reasons I got interested in that area. And then following my pathology residency, I finally went back and did a postdoc with Zena Werb at UCSF working on protease biochemistry and cell biology. So the answer as to why I'm into research is actually pretty easy—that's where I began.

Barry Bunin: Interesting. I read a *New York Times* interview with you awhile back where it mentioned that you'd done some cancer research and so

I'm familiar with your expertise in peptide chemistry and infectious disease. I'm also interested to hear how you initially got tenure at a real respected school like UCSF. Where did your own research start relative to where it is today?

Jim McKerrow: Well while I was doing my postdoc with Zena I was working on protease biochemistry. Her lab focus was on the role of proteases in macrophage cell function and innate immunity, how do macrophages move through tissue, how do they grade material they phagocytose, et cetera. And while I was working on those projects, I went to a seminar given by a parasitologist who's very well known in the Bay Area, actually internationally. His name is Don Heyneman. He is an epidemiologist who worked a lot in Africa on parasitic diseases and I was just stunned by his lecture. He was first a really stellar speaker, but he also talked about this organism which goes right through your skin and gets in your bloodstream and then develops. Here I had been looking at how single cells migrate through skin and he was saying, "Well there's this organism that has a thousand cells and it can go right through skin," and I thought: Hmm, sounds like a protease must be involved there, and so that's when I got hooked on the parasite protease side of things. And so when I finished my postdoc, I began working on how the schistosome parasite goes through skin and that was an initial interest and still is an interest of my lab. That in turn developed into a broader interest in parasitic infections and the pathogenesis of parasitic diseases. And along the way, the cancer side of it was because I had done a lot of work on proteases and because tumor cells, as well as the inflammatory cell response to tumors, generate a lot of proteolytic activity which allows tumors to invade and metastasize. I continued to be part of a program project group at UCSF with Zena, Lisa Coussens, Charlie Craik, Matt Bogyo from Stanford, and Jon Ellman from Berkeley looking at the role of proteases in cancer. There are actually some interesting parallels between the use of proteases by parasitic organisms to invade their host or to acquire nutrients from their host and tumor cell invasion of surrounding tissue and metastasis.

Barry Bunin: So within your current area of focus with the parasitic diseases, what's changed and what's evolved in your own group's research over time and also for the field as a whole perhaps there's some parallels...

Jim McKerrow: Yeah, taking first the local thing, my initial interest was in the basic biology of parasitic infections—host invasion, residence in the host, acquisition of nutrients, evasion of the host immune response—and my focus, as I said, was primarily on the role of proteases in these activities. But along the way, we began to broaden out from just looking at the schistosome parasite to other major neglected diseases, These would be diseases caused by parasites which affect hundreds of millions of people worldwide but for which there is little if any commercial interest in the pharmaceutical industry because these are diseases of poor people and poor regions of the world. Included in this group are diseases caused by single cell organisms like African trypanosomiasis; Chagas disease in Latin America, leishmaniasis, which is pretty much throughout the tropical world; amoebiasis, the diarrheal disease, malaria. So our focus, our interest was in the role of proteases in these different diseases. But particularly as we began to look at Chagas disease, which is caused by a single cell trypanosome, the major protease of that organism looked to be an exploitable drug target. Using my medical background, it was pretty clear that the current therapy for Chagas disease was inadequate, mostly due to the fact that there are severe side effects to current drugs and they have to be taken for 60–120 days! And so a kernel of thought began which was maybe we can at least come up with a new direction for drugs to treat Chagas disease by targeting this protease. Beginning in the late 1980s/early 1990s, I brought together a group of researchers from both UCSF and outside who covered a number of different areas ranging from chemistry with George Kenyon at UCSF and Bill Roush (who at that time was at University of Indiana), computer-based drug design with Fred Cohen and Tack Kuntz (and more recently Brian Shoichet), and protease chemistry with Charlie Craik, and parasitology with Phil Rosenthal. This group began asking the specific question: **Can we identify protease inhibitors that might be leads for drugs?** This work was supported primarily by the NIH. And then in 2002, as I was presenting some of this work at a local seminar, I was approached by **Herb and Marion Sandler** who had a foundation, and who commented, "we think what you're doing is just the kind of research that we consider important." And to make a long story short, they gave a substantial ongoing gift to the University to start what's now called the Sandler Center for Drug Discovery. The mission of the Center was to expand what we had been doing into a real drug discovery and development pipeline. While it was a big research challenge, it was something that we felt we could do and it was the logical evolution of our research. A lot of what we do now is directed towards drug discovery and development. As a result of their gift and also continued support from the NIH (and more recently The Gates Foundation), we've been able to bring in people from industry to UCSF to facilitate the drug development process. Industry was where the expertise was, so the Center at UCSF is now kind of a fusion between academic expertise and industrial expertise.

Barry Bunin: Thank you. So just turning a little toward CDD, how did you work with data and collaborations before CDD, and related why did your lab adopt CDD; what's been the value there?

Jim McKerrow: Right, initially when we began looking at potential targets for drug discovery and development, our focus was primarily on one type of target, cysteine proteases. At the time, a decade or more ago, the number of inhibitors that were available from any source, whether it be academic or biotech was pretty limited, particularly if we put the criteria on that these have to be inhibitors which don't just hit the enzyme but also can be used in tissue culture, can be used in animal models, have good solubility; reasonable pharmacokinetics and safety, there weren't very many. And so initially the data that we collected was something where any one of us would remember what we had tested in the last six months. There wasn't a need for a database. But over the last decade as our efforts ramped up and we moved into other disease areas, and as we developed high throughput screening assays where we could screen compounds in HTS/HCS microtiter format against biochemical targets or the actual parasites, the amount of data that came out of those expanded screens was beyond anything that any individual could remember. It really required a database that was searchable and easy to enter data into, and that's when we began to interface with CDD *because you guys had exactly what we needed*. This was part of an evolution as we transitioned from an academic consortium, a program project kind of group, into a more of a biotech fusion where we're working more like a company. We needed an "industrial strength" database.

Barry Bunin: Obviously we're ongoing and evolving continuously, but how has it played out for your own lab and for others as you've seen CDD develop over the years and what do you see as the next challenges for us as well?

Jim McKerrow: Well I think that the evolution of CDD, particularly for us, is that the organization, the company has been very receptive to developing the types of tools that we need, like being able to separate open source data from IP sensitive data. I think the challenge now, which you've hinted at, will be to better coordinate the efforts of the groups around the world that are in this arena because the funding is limited. One can't expect each of these groups to independently do what they want to do. It has to be more cooperative and collaborative and so the challenge is going to be: How do you coordinate efforts when you're dealing with different entities from different academic centers, biotech, industry, and international agencies, each of which may be under regulation and restraint by the funding agencies, venture capital, or stockholders that have supported them so far. That is going to be the next challenge.

CDD Spotlight Interviews: https://www.collaborativedrug.com/cdd-blog/tag/spotlight-interviews

There is an old story where the Buddha sits under a Bodhi tree and gets complete Enlightenment around 500 B.C.E., the so-called Bodhi Day. After leaving the comforts of a castle, experiencing suffering, and compassion for others, the Buddha came up with four noble truths, "This is stress... This is the origin of stress... This is the cessation of stress... This is the way leading to the cessation of stress...My heart, thus knowing, thus seeing, was released from the fermentation of sensuality, released from the fermentation of becoming, released from the fermentation of ignorance. With release, there was the knowledge, "Released." I discerned that "Birth is ended, the holy life fulfilled, the task done. There is nothing further for this world." From *Dvedhavitakka Sutta: Two Sorts of Thinking* translated from the Pali by Thanissaro Bhikkh. He could have stayed in that state forever. Instead, because of his compassion for others, he went back into the world and became a Bodhisattva to help others.

APPENDIX

Biographies

BARRY BUNIN, PhD
CEO & President

Barry Barry A. Bunin, PhD is the CEO of Collaborative Drug Discovery. Dr. Bunin has overseen $100 million in business transactions over the last two decades. Prior to CDD, he was an Entrepreneur in Residence with Eli Lilly & Co. Dr. Bunin is on a patent for Kyprolis™ (Carfilzomib for Injection)—a selective proteasome inhibitor that received accelerated FDA approval for the treatment of patients with multiple myeloma that was widely viewed as the centerpiece of Amgen's $10.4 Billion acquisition of Onyx Pharmaceuticals.

Dr. Bunin was the founding CEO, President, & CSO of Libraria (now Eidogen-Sertanty). At Libraria, he led a team that integrated exhaustive reaction capture (synthetic chemistry) with gene-family wide SAR capture (medicinal chemistry). On the scientific side, he co-authored "Chemoinformatics: Theory, Practice, and Products" (Springer-Verlag), a text that overviews modern chemoinformatics technologies, and "The Combinatorial Index" (Academic Press), a widely used text on high-throughput chemical synthesis. In the lab, Dr. Bunin did medicinal synthetic chemistry developing patented new chemotypes for protease inhibition at

Axys Pharmaceuticals (now Celera) and RGD mimics to inhibit GP-IIbIIIa at Genentech. Dr. Bunin received his B.A. from Columbia University and his PhD from UC Berkeley, where he synthesized and tested the initial 1,4-benzodiazepine libraries with Professor Jonathan Ellman.

KYLE RICHES
Head of Demand Generation

Kyle Riches II is a seasoned biotech marketing professional currently serving as Head of Demand Generation at Collaborative Drug Discovery, where he oversees marketing automation and sales enablement initiatives. With a strong foundation in sales development—having previously led teams at RaaStr—he brings a proven track record of building and managing high-performing sales organizations. He holds an Associate of Arts in General Studies and Business from Green River College, equipping him with a grounding in both operational strategy and team leadership.

JANICE DARLINGTON
Voice of CDD

Janice Darlington is the Voice of CDD at Collaborative Drug Discovery, where she leads customer communications and engagement initiatives. With over six years as a Customer Engagement Scientist at CDD, she brings deep scientific expertise and a passion for user success. Janice's extensive career in biotechnology includes senior research roles at Sanford Burnham Prebys, PharmAkea, and Bristol-Myers Squibb, where she played key roles in assay development, mechanistic studies, and cross-functional data integration.

Janice also brings a strong entrepreneurial spirit to her work, having co-founded and managed Altitude Trampoline Park Vista, where she led community marketing and events that elevated the park's brand presence in California. She holds a B.S. in Microbiology and Asian Studies from UC San Diego.

ERIC PUTNAM
Application Scientist

Eric Putnam is an Application Scientist at Collaborative Drug Discovery, where he supports customers in maximizing the value of the CDD Vault platform through technical expertise and scientific insight. He joined CDD in 2024, bringing over two decades of experience spanning biotechnology, software solutions, and applied fermentation science.

Prior to joining CDD, Eric served as a Solutions Engineer at Dotmatics, specializing in technical presales and scientific informatics. He previously spent over 13 years at Jacobs, where he advanced to Associate Scientist II, contributing to fermentation technology and applied biosciences. Earlier in his career, he held a key role as Fermentation Scientist at Fort Dodge Animal Health.

Eric holds a Master's in Technology Management from the University of Arizona.

MARIANA VASCHETTO, PhD
Head of Operations EMEA/LATAM

Dr. Vaschetto drives our global collaborations from CDD's new European Office. Previously she was the VP of Strategic Partnerships, Sales, and Marketing at Dotmatics and held similar executive roles at Perkin Elmer (previously CambridgeSoft). Mariana's broad experiences span across informatics, chemistry, physics, biology, and communication fields. She rose through the ranks at Accelrys from responsibilities for support to informatics applications to Chemoinformatics Software Product-Marketing Manager to Senior Product Manager for their complete product line (including data content and software). Mariana has a flair for cooking and knows which way the wind is blowing. Her PhD is in Chemistry.

JOHN MCCARTHY
Graphic Designer

John McCarthy is the Founder and Creative Director of JMC3 Design, a UK-based branding consultancy. With a background in leading large creative teams and working with major brands like the BBC, Tesco, and Dior, John now partners closely with startups and SMEs to craft purposeful visual identities.

At JMC3, John emphasizes clarity, collaboration, and consistency—approaching brand not just as design, but as a framework for decision-making and communication. He also leads Art of School, a specialist service supporting independent schools in elevating their brand presence.

CHARLIE WEATHERALL
Director of Customer Engagement

Charlie Weatherall is CDD's Director of Customer Engagement. He attributes his extreme customer dedication to two things. First, he is a true Southern gentleman who cares greatly about other people. Second, he started his career as an R&D end-user of multiple scientific applications and has never forgotten what it's like to "be the customer." Charlie has over two decades of experience working with industry-leading scientific software companies such as MDL (dating back to ISIS version 0.9 in '91), SciTegic, Accelrys, CambridgeSoft, and IDBS. He has traveled extensively installing software, providing training, and leading workshops at customer sites and conferences. Customers throughout the world (including Japan, Australia, Great Britain, France, New Jersey, and Illinois) have been known to alter their schedules to match Charlie's availability for visiting their sites. Charlie is the best customer advocate and trusted advisor for the CDD community.

KRISHNA DOLE
Chief Technology and Security Officer

Krishna Dole has over two decades of experience creating collaborative scientific software. He has developed software for researchers in drug discovery, population biology, geomorphology and phylogenetics. Krishna champions software engineering best practices, including test-driven development, security-focused design, and the original principles of the Agile Manifesto. He finds tremendous fulfillment working alongside CDD's team of domain experts and software engineers, who bring together deep scientific knowledge, technical excellence, and a shared commitment to advancing drug discovery research. Krishna holds a double BA in Biology and Earth Sciences/Environmental Studies from UC Santa Cruz.

KELLAN GREGORY
Director of Product

Kellan Gregory is CDD's Director of Product. He has a degree in Chemical Engineering with an emphasis in Biotechnology from Tufts University. Kellan was co-author in an HIV study with Dr. Paul A. Volberding, and interned with BioRad and Libraria. He has not written two books and founded two companies like Barry Bunin, PhD. Unlike Sylvia Ernst, PhD, he did not help with commercial introductions of both Beilstein and Pipeline Pilot. And he's at least 50 publications behind Alex Clark, PhD. But you know what? If you want a product expert with 20 years of experience wrangling data, the strongest people skills and the ability to tell you straight up what will and will not work, Kellan Gregory—with his growing team of specialists in the product group—is your best collaborator. Kellan and his team creates practical specifications for our product enhancements by prioritizing thousands of requests from hundreds of CDD customers.

ABRAHAM WANG
Head of Marketing

Abe Wang is the Head of Marketing at Collaborative Drug Discovery, where he leads strategic initiatives to elevate the CDD Vault brand and drive global engagement within the life sciences community. Since joining CDD in 2019, Abe has been instrumental in expanding brand awareness and digital marketing initiatives, contributing to the company's continued growth as a leading SaaS informatics platform for scientific data management.

Abe brings over a decade of experience at the intersection of science and marketing. Prior to CDD, he held marketing leadership roles at Thermo Fisher Scientific and Affymetrix, where he specialized in market development, demand generation, and product strategy for life science technologies. His foundation in scientific research was established at SRI International, where he contributed to preclinical drug development and translational research for over seven years.

Abe holds an MBA from Santa Clara University and a Bachelor of Science in Human Biology from the University of Toronto.

SYLVIA ERNST, PhD
Sr. Manager, Commercial Operations

Dr. Sylvia Ernst is a respected figure in the business realm of chemical information technology, bringing her global business experience to the leadership team at CDD.

Since joining CDD in 2007, Dr. Ernst has played a pivotal role in shaping the company's success. She is driven by the steadfast belief that leveraging technology for collaboration, paired with robust data management, can greatly expedite the discovery of new therapies. Her work ranges from customer-facing roles to operational ones, facilitating strategic partnerships and negotiating impactful contracts. In her past roles at Accelrys, SciTegic, Elsevier MDL, MDL Information Systems, and Beilstein, Dr. Ernst honed her expertise in customer-centric and management roles. She earned her PhD in Chemistry at J. W. Goethe University Frankfurt in Germany under the esteemed Prof. Dr. Wolfgang Kaim. With over 20 scientific publications to her name, Dr. Ernst's early research focused on the innovative use of computational chemistry to discover new catalysts for generating hydrogen from water and sunlight.

SALIMA ISMAYILOVA, PhD
Director of Business Development

Salima Ismayilova is a Director of Business Development at Collaborative Drug Discovery, where she helps life science organizations accelerate research through strategic adoption of the CDD Vault informatics platform. Since joining CDD in 2022, she has brought a uniquely global perspective to commercial growth, shaped by her cross-industry experience in agriculture, technology, and scientific research.

Prior to CDD, Salima held several leadership roles in the farming and construction equipment industries. During her nine years at John Deere, she led senior sales efforts focused on technical project delivery and market expansion across Azerbaijan. She later joined Guntert & Zimmerman, where she enabled international sales and helped drive revenue growth.

Salima earned her PhD in Botany/Plant Biology from the Azerbaijan National Academy of Sciences, conducting research in a Bioactive Compounds lab with a focus on aromatic compounds in oak moss lichens. She also holds an MBA in International Business & Negotiations from Florida International University, graduating with distinction.

JACOB BLOOM, PhD
Senior Developer

Jacob Bloom is a Senior Software and Infrastructure Engineer at Collaborative Drug Discovery, where he has played a pivotal role in developing and scaling the CDD Vault platform since 2016. His work focuses on backend software architecture, infrastructure optimization, and enabling robust, secure systems that support global scientific collaboration.

Prior to joining CDD full-time, Jacob contributed to scientific software development at Chemical Semantics, Inc., where he led web portal design for computational data sharing and advanced the integration of semantic technologies. His early career also includes multiple academic teaching roles in chemistry and physics at institutions including Texas A&M University, University of Georgia, and New College of Florida.

Jacob holds a PhD in Physical Chemistry from Texas A&M University, where his research centered on modeling non-covalent interactions and π-systems. He also earned a Master's degree in Physical Chemistry from the University of Georgia.

JONATHAN BISSON, PhD
Senior Developer

Jonathan Bisson is a Senior Software Engineer at Collaborative Drug Discovery, where he combines his deep domain expertise in natural product research with advanced software engineering to enhance the CDD Vault platform. Since joining CDD in 2021, he has contributed to infrastructure automation, cheminformatics tooling, and the integration of AI technologies including large language models (LLMs) and deep learning pipelines.

Jonathan brings more than a decade of experience at the intersection of computational chemistry, pharmacognosy, and software development. Before CDD, he served as a Research Assistant Professor at the University of Illinois at Chicago, where he led NIH-funded efforts in antibiotic discovery and built custom tools for mining biomedical literature, managing high-through-put instrumentation, and advancing natural product research. He was also the Lead Architect of the redesigned NAPRALERT database, one of the world's most comprehensive repositories of natural products data. He also co-created LOTUS, the largest public-domain and collaborative natural products information source on Wikidata.

He holds a PhD in Interface Chemistry-Biology from the Université de Bordeaux, graduating cum laude, and has been a driving force in open data, ontology design, and full-stack systems for research informatics.

PETER GEDECK, PhD
Research Informatics Senior Scientist

Peter Gedeck is a Research Informatics Senior Scientist at Collaborative Drug Discovery, where he designs and develops production-quality software solutions that support scientists in accelerating drug discovery. Since joining CDD in 2017, Peter has contributed his cheminformatics expertise and computational chemistry insights to the ongoing enhancement of the CDD Vault platform.

Peter brings more than two decades of industry experience, including nearly 20 years at the Novartis Institutes for BioMedical Research (NIBR). There, he held successive roles in computational chemistry and cheminformatics, supporting research projects across several disease areas including respiratory and tropical diseases. He played a pivotal role in strategic software development, including the FOCUS scientific portal, and led cross-functional teams spanning multiple continents.

Peter earned his PhD in Chemistry from FAU Erlangen-Nürnberg, where he specialized in quantum mechanical modeling and spectroscopic studies of electron transfer processes. In addition to his work at CDD, Peter teaches data science at the School of Data Science at the University of Virginia.

JASON HARRIS, PhD
Data Wrangler

Dr. Jason Harris has a broad background in biological experimentation, biophysical simulation, chem/bio-informatics, software development, business development, grants and compliance. Jason works at the interface of science and technology, helping to make data shareable and reusable. Jason created and maintained ScrubChem.org which is a digital curation of the NIH PubChem-Bioassay. ScrubChem allows for building the large datasets that are needed to move forward the state of art for bioassay design and molecular modeling. ScrubChem restores interoperability and re-use of 1.2 million public bioassays, 2.3 million chemicals, and over 1.5 billion records of data. Jason works at CDD on several tools to improve the management and reuse of biomedical data, he wears many hats and solves many problems. Jason earned his PhD from University of Tennessee & Oak Ridge National Laboratory, and postdoc at University of North Carolina and the EPA.

ALEX CLARK, PhD
Research Scientist

Alex Clark has been building cheminformatics and computational drug discovery software products since the early 2000s, after spending a decade learning to be a scientist the hard way: running reactions at the lab bench. Since switching to informatics full time, he has worked on most areas of contemporary chemistry software, whether it be 2D or 3D, large scale (cheminformatics) or small scale (quantum chemistry), big molecules (proteins) or small molecules (drugs). He has a persistent interest in the interface between scientists and software, which continues to challenge our industry, and has done much work on visualization techniques, electronic lab notebooks, and reimagining traditional software for new platforms like mobile, cloud and web.

He publishes regularly in the scientific literature, and has maintained a large number of collaborations since becoming an entrepreneur in 2010. Whenever he is not working on projects for CDD, he continues to maintain his own company, Molecular Materials Informatics, relentlessly experimenting with all manner of future-oriented chemistry software ideas, methods and products. Alex completed his doctorate at the University of Auckland, New Zealand, in 1999, and subsequently relocated to North America. Read Alex's expanded bio including publications with citations and other resources.

JOE CLIFFORD, MFA
Writer, Musician, and Painter

After spending the 1990s as a homeless heroin addict in San Francisco, Joe Clifford got off the streets and turned his life around. He earned his MFA from Florida International University in 2008, before returning to the Bay Area. His memoir, Junkie Love, chronicles his battle with drugs and was first published in 2010 and re-released in 2018. He is the author of the award-winning Jay Porter Thriller Series, as well as several standalones including *The One That Got Away, The Lakehouse, The Shadow People, Say My Name, All Who Wander and A Moth to Flame.*

His bestselling Jay Porter Thriller Series (Oceanview Publishing) has received rave reviews from *Publishers Weekly, Library Journal,* and the *San Francisco Chronicle*, among many others. Joe is also editor of *Trouble in the Heartland: Crime Stories Based on the Songs of Bruce Springsteen* and *Just to Watch Him Die: Crime Fiction Inspired by the Songs of Johnny Cash*. Currently Joe teaches online writing courses for FIU, as well as around the country at various conferences and retreats.

CHRIS WALLER, PhD
Strategist, Collaborator

Chris Waller is the Chief Strategist at Quantori. Chris Waller was the VP and Chief Scientist at EPAM Systems, specializing in life sciences informatics strategy and technology development. Previously, he served as Executive Director at a large global pharmaceutical organization, where he led scientific modeling platforms and real-world evidence initiatives. His industry experience includes leadership roles in enterprise architecture and healthcare informatics, where he advanced the use of electronic medical records and personal health record data to inform research and development efforts. He has been instrumental in pre-competitive technology development, serving as a founding evangelist of the Pistoia Alliance and leading strategic initiatives in chemistry informatics. Chris has extensive experience managing large-scale technology implementations, including leading a team of over 100 scientists and engineers across global operations. His contributions to open-source development and pre-competitive collaboration have helped shape industry standards in pharmaceutical research informatics. Chris holds advanced degrees in life sciences informatics and has over 30 years of experience in strategic planning and organizational development.

PAT WALTERS, PhD
Chief Scientist, Collaborator

Pat Walters, PhD is Chief Scientist at OpenADMET, focusing on open, reproducible ADMET modeling and the application of machine learning to small molecule drug discovery. Previously, he served as Chief Data Officer and Senior Vice President of Computation at Relay Therapeutics, where he led teams integrating computational chemistry and data science into structure-based drug design programs. Earlier, he was Principal Research Fellow at Vertex Pharmaceuticals, managing global teams in molecular modeling, cheminformatics, and bioinformatics for computational drug discovery. Dr. Walters holds a PhD in Organic Chemistry from the University of Arizona and a BS in Chemistry from the University of California, Santa Barbara. He is recognized for advancing data-driven approaches to predictive modeling and for promoting open collaboration in the computational chemistry community.

DR. JAMES MCKERROW, PhD
CDD Spotlight Interview

James McKerrow, MD, PhD was the director of the Sandler Center for Drug Discovery and Vice-Chair for Research Affairs at the University of California, San Francisco (UCSF). Later, Dr. McKerrow was Dean of the

School of Pharmacy and Director of the Center for Discovery and Innovation in Parasitic Diseases at UCSD. His group utilizes techniques from a variety of disciplines to develop new drugs for parasitic neglected diseases to clinical trials. As CDD's first major customer, Dr. McKerrow guided us in our early days and introduced us to the Gates Foundation, making him perfect for our first CDD Spotlight.

AFTERWORD

Entrepreneurial Lessons from
Dr. Alejandro Zaffaroni

FOUNDER'S NOTE:

In India, I saw that the chemists were just as talented and, of course, working at a fraction of the costs as in the USA. I wanted to set up my own combinatorial chemistry company in India, which I called Libraria—for libraries of molecules and libraries of data. It was around the time the internet was first skyrocketing, so we decided to pivot to databases. When founding Libraria, I wanted to work with Dr. Zaffaroni, a well-respected entrepreneur who had a number of break-throughs under his belt (the birth control pill, the patch, the GeneChip). Dr. Mark Gallop introduced me to Dr. Zaffaroni, and I had the honor, privilege, and pressure of having him as my business mentor.

D r. Zaffaroni was appropriately respected. He had a charisma and gravitas that captured everyone's attention when he walked into a room. He was generous with his time, meeting with me on weekends. I shared two anec-dotes in an obituary when he passed away. One story was about how he would drive us over to meet with the CEOs of his other "Zaffaroni Companies," even though he could have had them come to his top floor office at ALZA. I asked Dr. Zaffaroni why we were going there, given his time was so valuable and he said, "Everyone's time is valuable." Another time Dr. Z was advocating for his idea (natural product libraries). When I timidly suggested I had another idea, he looked at me and said, "That's good. Otherwise, what would I need you for?" He meant it as a compliment and a joke.